Essays from Earth's History

DANKMAR BOSSE

ESSAYS FROM EARTH´S HISTORY

ADDITIONAL STUDIES CONCERNING THE MUTUAL EVOLUTION OF EARTH AND HUMANITY

Formative Forces Publishing
2021

Translated by Jannebeth Röell and James Lee and reworked together with the author.

Formative Forces Publishing

formativeforces@icloud.com

ISBN Print: 978-1-7365513-1-8

ISBN eBook: 978-1-7365513-0-1

Front Cover: a granite specimen with large orthoclase-feldspar crystals (red), in between small quartz spheres (gray), and a few mica crystals (black). See Fig. 76, page 91.

Frontispiece: Minerals in a typical granite sample. See Fig. 106, page 134.

Library of Congress Control Number: 2021902004

Printed in the United States of America

Contents

Seek in your own being:
And you find the world;
Seek in the world's ruling powers:
And you find yourself;
Take note of the pendulum-swing
Between self and world:
And there will be revealed to you
The human's cosmic being;
The cosmic human being.

– RUDOLF STEINER (CW 40, February 27, 1919)

And the teaching concerning the human being,
As formed from the forces of the Earth,
Is geology.

– RUDOLF STEINER (CW 149, December 12, 1913)

Introduction

This book is written for all readers interested in the Earth sciences, geology, Goethean Science, and Rudolf Steiner's anthroposophy. It builds and adds to what I wrote about in *The Mutual Evolution of Earth and Humanity: Sketch of a Geology and Paleontology of the Living Earth.* There I compare and contrast the research results of geology and paleontology with those of Anthroposophical Spiritual Science. The Earth's history is an encompassing process that has sense-perceptible and suprasensible sides. The findings of spiritual-scientific research can be tested on nature's phenomena and the Goethean path of knowledge. This way leads through the observations to concepts and ideas which are experienced as true. I have found that geological connections, down to the details of rock, fossil, and landscape, provide an astonishingly faithful image of the human being's suprasensible development, as Rudolf Steiner describes it. In *Essays from Earth's History,* I explore how these topics and methods intersect with our modern lives.

My destiny brought me to the natural sciences and anthroposophical knowledge. In 1940, I was born in Weimar, Germany, the city of Goethe. I grew up there and started collecting stones at the age of twelve. Since then, it was clear that I would become a geologist. I read the philosophical books of Guenther Wachsmuth (1950), Walter Cloos (1958), and Helmut Knauer (1961) even before my geological studies. At the Freiberg University of Mining and Technology in the Ore Mountains in Central Europe, I received focused training in geology, mineralogy, tectonics, geophysics. I wrote my thesis in 1968 in paleontology under Professor Arno Hermann Müller. From the beginning, my goal was to examine the mutual relations between anthroposophy and geology critically.

A third field was added to these. As a citizen of Weimar, I had been reading Goethe's geological writings from a young age. A synopsis of his basic ideas, which I presented in a first manuscript in 1976, was published in 1994 under the title "Die Metamorphose des Granits" (The Metamorphosis of Granite). I saw an important starting point for a new idea in Goethe's thoughts, the evolution and development of the Earth as a living being. Goethe's hunches can become a link between the spiritual side of evolution, as presented by Rudolf Steiner, and the sense-perceptible side described by the modern geosciences. The geosciences, anthroposophy, and Goetheanism became the three points of departure for this work – both methodologically and in content. I have had the opportunity for the first time in 35 years to devote a large part of my working time to these topics.

The company for which I worked is dedicated to its customers. During the 21 years of activity at the Central Geological Institute in Berlin, I worked mainly on questions about the regional geology and tectonics of Central Europe and worldwide comparisons, and I had first-rate specialty libraries at my disposal. I had the opportunity to study a wide variety of geoscientific research.

I was able to scrutinize Rudolf Steiner's representations and to compare them with that research. I was also able to continue working while studying anthroposophy. Additionally, there were numerous excursions during which thousands of handpieces were struck and rock sequences compared. The contrasting ideas were again and again brought to the images of the rocks and landscapes. A comprehensive view slowly developed from the details, but it deviates from the concepts we had learned in some respects. I want to thank very much all of the friends who helped me through the years, through lectures and seminars, in geology classes at Waldorf schools, on excursions, and in conversations.

1. The Application of Healing Stones

Pathways of Oral and Historical Traditions

For many years now, it has been possible to do a profitable business with healing stones. Some of the twenty or so minerals known from legend were rare and sold out quickly. As a result, several hundred minerals with similar colors were soon offered as "healing stones." For example, instead of the very rare topaz, there were pieces of smuggled amethyst that became a translucent yellow when exposed to radiation. Suppose one asks about the basis of their effects. In that case, the answer is, on the one hand, personal experience and the other traditions founded on unsteady ground. The aim here is to investigate this subject and take a closer look at it with spiritual science and medical knowledge.

In early antiquity, people could still perceive the effect of gemstones, but already by Greek times, tradition had become predominant. The oldest source describing the minerals and their characteristics is the fragment *About the Stones of Theophrastus* (372 - 287 BC; written about 314 BC). He was a student of Aristotle (384 - 323 BC) and made a distinction between Earth, metals, stones, and salts according to Empedocles in his theory of the elements (490 - 430 BC). Pliny in Rome (23 – 79 AD) summarized the ancient knowledge in his *Natural History* (around 70 AD). He also reviewed what was known about medicinal effects, especially metals, but already in part without understanding it himself. Galen (approximately 131 - 201 AD) also described the minerals used medicinally, whose names and experiences he adopted from the Greeks. Democritus, Theophrastus, Zoroaster, and Orpheus described the healing powers of gemstones and their colors. Orpheus had his knowledge from Hermes Trismegistus (Menes, around 3,000 BC), the founder of Egyptian culture, who had revealed the effects of three times nine gems to Orpheus.

In the Arabic writings from the 6th century onwards, numerous transcriptions were created from Aristotelian traditions. The miraculous effects and powers of minerals were no longer understood by that time. They were mixed with external characteristics by which they are distinguished, such as the place where they were found, color, and density. For example, as found in the *Stone Book of Aristotle* by Hunain Ibn Ishak (around 830) or the Arabic doctor Avicenna (980 - 1037). This trend continued into the monastic schools of the Middle Ages, where the Arabic writings were translated into Latin. Most notably, Albertus Magnus (Albert of Cologne, 1207 - 1280) collected the knowledge of his time in five books about stones around 1260. What was still known of the minerals, especially of the compounds, can be traced back to him in the later Middle Ages, also the names of the gemstones in Luther's Bible translation. Albertus describes the formation of the ores, metals, and rock-forming minerals through the planets' forces and fixed stars, bringing about the crystal forms. Supernatural forces are not doubted, but he says that they can only rarely be experienced. Therefore, in the beginning, he refers to numerous sources, among them contemporary alchemists such as Joseph and others; the Arabs such as Avicenna; the Romans Pliny, Galen, and Dioscorides; and even more ancient ones such as Aaron and Hermes Trismegistus. He describes 96 minerals and lists everything that has been handed down from antiquity about the effects of metals, stones, and gemstones on the human being. Another well-known tradition is based on the visions of Hildegard von Bingen (1098 - 1179), which she described in her "*Physica*."

The healing effect of the gemstones was experienced almost exclusively through their colors. For example, there was a connection between a red stone and blood diseases. Thus, in books of antiquity, the stones were distinguished and named primarily based on color properties. Only through the Arabs were other external characteristics added, such as hardness, transparency, and luster. This situation was still the case throughout the Middle Ages and into the modern age, as was the division of substances according to the four Greek elements of fire, air, water, and Earth. In Weimar, Goethe still arranged his collection of minerals systematically according to these Greek elements. But these characteristics were no longer sufficient because of increasing technological processing of the minerals. At the end of the 18th-century, minerals and gemstones began to be differentiated through chemical analysis by the likes of Berzelius and given names. The modern chemical system was founded at the Freiberg University of Mining and Technology by Abraham Gottlob Werner (1749 - 1817) and M. H. Klaproth (1795 - 1815). Today, fine chemical, atomic spectroscopic, and structural analysis are the primary basis for naming.

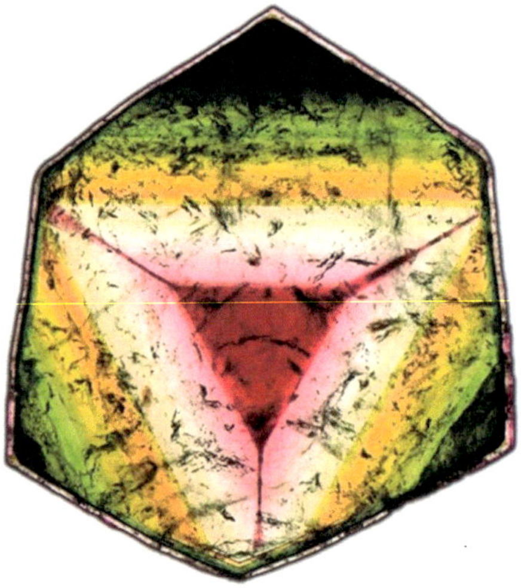

Fig. 1: Tourmaline (Liddicoatite) Madagascar

This naming mode is demonstrated with tourmaline (Fig. 1), whose multi-color variety fills the entire color circle. Microchemical analysis confirms that metallic substances give the tourmaline its colors. Colorless zones are free of them. The iron produces bivalent green and trivalent red, brown, and yellow color nuances, which become black when intensified. There are also the colors of the iron-related metals: green from chromium, vanadium, and titanium; blue and light red from manganese; brown and yellow from titanium; blue, from copper, rarely also from iron. Many black Schorl crystals have been shown to have color when thinned to a tenth of a millimeter thickness. The different tourmaline minerals are mainly distinguished by their chemistry, after which they are named. Elements are decisive for the naming, for example: Ca + Li: Liddicoatite; Na + Li: Elbaite; Ca + Mg: Uvit; Na + Mg: Dravit; Na + Fe: Schorl, Olenite or Fluor-buergerite, etc. Paul Rustemeyer writes about the color formation: "In white light, the tourmaline crystals appear colorful because some of the incident light is absorbed. The remainder appears in the human eye as a complementary color." [1] It explains different possibilities of absorption by the electrons' charge processes in the molecular lattice, leaving open how the charge of uncolored electrons can determine a mineral's colors.

Gemstone lovers and collectors live in a completely different world. They still name Tourmalines after their colors, for example: red-Rubellite (Latin rubelos = red); green-Verdelite (french vert = green); blue-Indigolite (bluestone); colorless-Achroit (not colored). So it would be enough to say "red tourmaline" instead of rubellite, although a Latin name sounds more scientific and feigns greater accuracy. Both ways of naming are used together – chemical and color – in working back from the color to the chemistry. For example, when a multicolored tourmaline without chemical analysis is named Elbaite, and so on. This method is not clear and has led to many names and confusion.

Another issue connected with the direct healing effect is the relation of gemstones to the whole human being. One aspect is the relation of the physical body's twelve external aspects to gemstones since they are related to the zodiac signs. Some fragmentary notes and early lectures by Rudolf Steiner concern gemstones and metals connected with Earth's evolution and humanity. "Minerals have a group soul in the higher spiritual world. Now, if stones have souls, you will also understand how a moral relationship can arise. [...] The body of a stone has neither desire nor urge; therefore, it presents the human being with an ideal through which our urges may be spiritualized. And in the distant future [...], the human being will be like a diamond. [...] Chaste and virginal, the stone preserves this Creator Word in the depths of physical existence. [...] There was also a time when gold was not yet solid; this was the case when the Sun and Earth were still united. When the Sun left the Earth (in the Hyperborean epoch), traces of this finer substance were still evident. Like all metals, it flowed and then formed veins of gold in the rock. Going back even further, we come to a time when gemstones also solidified. [...] The astral body had the desire to perceive something, to see, and then it acted on the etheric body and formed the etheric eye out of it; only later was the physical eye separated. An etheric counter-image arose in the mineral kingdom, and this desireless counter-image is chrysolite (the gemstone variety of olivine). Thus there is indeed an intimate connection between human vision and chrysolite. This example explains why the occultist uses stones for special purposes. He or she feels a sympathy between vision and chrysolite and knows how to use them to treat certain eye diseases." [2]

A week later, he voiced a similar thought in another lecture: "There is a fundamental experience in the mineral world. Let us take the rock crystal, glittering with light. When one looks at this, one says to oneself: In a certain way, this represents an ideal for the human being. The stone is physical material in the same way the physical body is physical material. But there is a future perspective to which the occult teacher leads the pupil. The human being of today is still penetrated by instincts and desires, by passions. [...] He says to himself: the animal nature of the human being will gradually be cleansed and purified to a stage where the human body can stand before us just as inwardly chaste and free of desire as the mineral that craves nothing, in which no wish is stirred by what comes near it. Chaste and pure is the inner material nature of the mineral. This innocence and purity is the feeling that must permeate the pupil on gazing at the mineral world. These feelings vary as the mineral world shows itself in different forms and colors, but the fundamental experience which permeates the mineral kingdom is chastity. [...] As a piece of ice freezes out of water, so are the physical organs formed out of the finer etheric body. The physical organs were formed within man while outside the Earth became solid. In every age, forming a human organ took place parallel with forming a particular configuration outside in nature. [...] There is also a relationship between the onyx and the organ of hearing. The onyx stands in a remarkable relation to the oscillations of a person's ego-life. [...] A further occult relationship exists between the organ of taste and the topaz (sense of taste), the sense of smell and jasper, the sense of skin as the sense of warmth, and the carnelian (sense of touch), the productive power of imagination and the carbuncle. This stone was used as the symbol for a productive power of imagination, which arose in the human being at the same time as the carbuncle in nature. The diamond in the Saturn epoch formed the germ of the physical body." [3]

A page from Rudolf Steiner's notebook related to the same lecture records how the physical body of the human being was formed in the elementary world in connection with several gemstones:

"Elem[ental] world:

Sardonyx	Head
Carnelian	Neck
Topaz	Arms
Chalcedony	Chest
Jasper	Heart
Emerald	Belly
Beryl	Kidneys
Carnelian	Feeling
Amethyst	Reproductive organs

Hyacinth	Hips
Chrysoprase	Knees
Crystal	Thighs
Sapphire	Feet” [4]

Another notebook by Rudolf Steiner from 1906 lists relationships between the gems, the zodiac signs, and the planets with the human being's developmental periods. The attributions correspond with those of Agrippa von Nettesheim (1533). Other authors of today give somewhat different information. With the names that Rudolf Steiner used, one could assume that it would be the common name used today. The similarity of Agrippa and Rudolf Steiner’s use of medieval names such as carbuncle, hyacinth, sardonyx, and crystal, suggests that the words refer to corresponding color qualities, not to minerals that are determined by chemistry in the modern sense.

There are 36 places in the bible where individual gemstones are mentioned. Two similarly translated lists of twelve gems are found at the beginning of the Old Testament, where Moses (around 1500 BC) is instructed to make a breastplate for the Egyptian high priest Aaron. At the end of the New Testament, John in the Book of Revelation describes New Jerusalem (around 100 AD), with its walls built of jasper and its foundation stones prepared from twelve gemstones. The names were translated from Hebrew into Greek (Septuagint) in ancient times and Latin since Luther’s time. In the spirit of Rudolf Steiner’s thoughts mentioned at the beginning, I see the spiritual effect of gemstones above all in the fact that they can be a picture for us, a silent summons to purify the soul. Suppose the gates of New Jerusalem in the Apocalypse are understood as the purified senses of the future human being’s spiritual body in the Jupiter period. In that case, they will represent a developmental goal that shines a light on us from the Earth’s oldest past.

There are also other translations or classifications whose significant differences are mainly related to the gems’ names, some of which have a long history. Lüschen [5] thoroughly researched the change in the meaning of gemstone names since antiquity by meticulously comparing historical texts, mineral characteristics, their colors, the location where the gemstones were found, and how language has evolved since that time. In this way, he has attempted to determine which modern and ancient gemstone names belong to which minerals. This question is important because, in early antiquity, the gemstones’ effect could sometimes still be experienced, but generally not anymore today. Nowadays, the published books on healing stones connect to this old tradition without considering the many name changes.

Nevertheless, in most cases, this practice has led to great misconceptions. Firstly, in ancient times the minerals were named after the spiritual experience of their colors, but since the 18th century, they have been named increasingly after their chemistry. In the past, many substantially different minerals were thus grouped under one name. Secondly, especially in the Middle Ages, names were transferred to individual group names or other minerals because even then, the traditions of antiquity had been lost. Thirdly, the names themselves have often changed and shifted considerably due to dialects, translations, and changes in meaning. Only the trends for the most important gems can be traced in a shortened form in what follows. Here, I order the minerals by color, following their ancient names.

Red gemstones were consistently called carbuncles or carbunculus by Pliny and also by Alabandicus. The name comes from the Greek for small coals, with a color similar to glowing coals, or Greek anthrax, a lump of burning coal. Pliny writes that numerous difficult to distinguish minerals are designated this way; garnets, pyrope and almandine, ruby, and carnelian (a red chalcedony), can be found among them. Albertus distinguishes three types: Balagius (red spinel), Granatus, and Rubinus, mainly known in the Middle Ages by the name Carbuncle, derived from blood colors.

Green gemstones can sometimes be distinguished more clearly. In the case of beryl and heliotrope, the ancient names correspond approximately to today's minerals. In ancient times, emerald (Greek and Sanskrit) referred to various minerals such as malachite, chrysocolla, nephrite, or the green varieties of fluorspar. It was not until the 18th century that it was distinguished from beryl by chemical analysis. Chrysoprase most likely belonged to the emerald of antiquity; today's name comes from the Middle Ages. Jasper comes from the Greek for a green, translucent stone; the present name, however, refers to an opaque chalcedony. In the Bible and Pliny, it is called the noblest stone. Most probably, it is not the stone with that name today. From the Greek ophthalmos = eye, the opal would also be a possibility, based on a description of the most beautiful of the 12 stones in the imperial crown of Otto I. Today the gemstones on such crowns are only referred to as jewelry. In ancient times, however, they were a sign of the imaginative vision of their wearers.

Blue gemstones were named from the Indian sapphire and were called yakut or blue hyacinth by the Arabs in the Middle Ages. In antiquity, this mainly referred to lapis lazuli or glazed stone, in the Middle Ages, and blue corundum. Only the amethyst corresponds to ancient use, where it means in Greek "not drunk," and is derived from the color of a strongly diluted red wine.

Yellow gemstones were called chrysolite in ancient times from the Greek gold stone or sunstone because it was translucent and shone like gold. Today the light green translucent opal is so named. The interpretation remains uncertain; it could be topaz, citrine, or the yellow varieties of fluorspar, beryl, or corundum. In antiquity, topaz was mainly called yellow, and also, many different colors; fluorspar was probably meant, which can be deduced from the hardness data. The Lynkurer, which became "Luchsstein" in the Middle Ages, was a yellow stone with many interpretations such as amber, hyacinth, or fossil belemnites.

Multicolored gemstones were called onyx. The current importance of black and white banded chalcedony has developed since the Middle Ages. In antiquity, one meant color-banded and speckled stones like agates or alabaster. Until the 18th century, agates were stones that exhibited colored, pictorial patterns, independent of the mineral substance. Today's agates are called sardonyx. The yellow-red Sarder – today a rather brown Chalcedony – may mainly correspond to a mineral named after the city in Asia Minor – presumably an ancient trading place.

In Greek, a diamond, demant, or adamas was an exceptionally hardened, artistically forged helmet. Pliny described it as indestructible and said the anvil and hammer would be more likely to shatter when striking it. It was also said to have magnetic properties. In antiquity, diamonds were probably unknown; today's term dates back to the 18th century.

The question of healing stones

It is possible to find numerous contradictory applications when comparing several modern books about healing stones (especially books by Michael Gienger [6]). A diffused field of effects is usually given without substantiation since they probably originate from non-transferable, individual experiences. Electromagnetic and light radiations are described, which would emanate from the stones, and resonances caused by biophotons in the cells. Physical terms are used to explain phenomena that are not physically detectable. A distinction is made between intuitive, energetic, empirical, and radiosensitive stone healings. I see the following challenges when what is described above is correlated with traditional applications:

1. Differences in gemstones' effectiveness with bodily changes in the human being since antiquity are not considered.
2. Modern chemically-based mineral descriptions are used.
3. The identity of traditional gemstone names is lost: they used to be given according to color, but today according to their chemistry.
4. Individual minerals are used as healing stones even though the names refer to traditions where the names of color groups stood for minerals with very different chemistry.
5. A discontinuity in tradition between antiquity and today, mainly between Pliny and Albertus; therefore, it is impossible to refer to traditions.
6. It is about individually different effects according to experience or capacities on which no rational therapy can be built – perhaps also auto-suggestive effects, similar to placebo effects.
7. No attention is paid to the difference in effect between an external application and the internal use of a pharmaceutically processed substance.

Rudolf Steiner has not given a conclusive answer to the question of the external effects of gemstones. In answer to a question asked after a lecture, he said: "There is a question that is difficult to answer. [...] How does the wearing of pearls and gemstones work upon individual organs? Certainly, there is an effect, but the effect can only be judged when one can look into the spiritual world; the effect has to be judged individually. It can quite well be said, for example, Sapphire works upon a certain temperament, upon a choleric temperament, but only in an individual case. There certainly are effects, but to answer the question completely, one would have to enter into deeper things than is possible today." [7] Such an answer is entirely understandable after what has been presented in this chapter. Rudolf Steiner goes on to describe the characteristics of a similar healing effect. He describes how "a person can have a healing effect through his will on the solid, fluid, airy and warm areas of his body; by giving stimulation to the astral body through exercises, [...] which he has performed, which has somewhat strengthened his will, which can be strengthened; [they] have acted on his astral body. The astral body has worked upon his etheric body. The etheric body has worked in a healing way on his physical body and has been able to adjust to nullify the trouble that is not a deep, organic one. [...] Out of our own astral body, without the patient exerting his own will, we can influence our own etheric body in such a way that our own etheric body works upon the etheric body of the patient in the same way as the astral body worked previously. That is healing magnetism. [...] The reverse is the case in therapy where medicaments are used. When you give medicaments, you introduce into the physical body substances that work partly upon the inner forces and partly upon the physical body's rhythm to influence the patient's etheric body. The recovery always proceeds from the etheric body. You influence the etheric body from the astral body, which is psychological healing to which magnetic healing belongs [...]. Or you have rational therapy, which starts with medical interventions that proceed from the physical body and pass into the etheric body. Always, however, the healing proceeds from the etheric body." [8]

Four years earlier, Rudolf Steiner asked in a lecture: "For what in the whole of nature's immensity is really beyond the human? Nothing, indeed. All that is external to our being in those portions of the world we can observe has been expelled or removed from us in the course of human evolution. Humankind has had to pass through developmental stages only possible because certain essential processes take their course in the outer world. It is thus enabled to take certain other processes into human beings for their use. So that there is always a complementary polarity and kinship between certain external and certain internal processes." [9]

This is a revised text of the first publication in Germany: Bosse, D.: Heilsteien auflegen Der Merkurstab, Issue 4, 2005, p.272-279.

2. Why do Mineral Remedies Have an Effect on Human Beings?

A question can arise about the relation between mineral substances and human evolution if one considers that the contrast in nature between humans and minerals is the most significant polarity. But a further question arises about the connection between these two kingdoms of nature when mineral substances are used as medicine. Viewed from today's geoscientific models, hardly any connection can be found between humans and minerals. The origin of the living nature kingdoms is described in these models as the random combination of dead substances by chemical processes. Yet, there are many arguments to the contrary; for example, all life processes are determined from within by an ordered time organism. In contrast, chemical processes occur immediately when the necessary conditions are present. And the models do not explain the origin of many substances that are only present on Earth.

In complete contrast to such models, Rudolf Steiner describes in his *An Outline of Occult Science* [10] how all beings of the natural kingdoms – including the minerals – emerged in the beginning from the spiritual human germ and developed and condensed through the elemental states of heat, air, and water to solidity. Scientific research rightly sees its task not in explaining such questions but rather in extracting the given substances, processing them, and transforming them into new materials with specific properties to be used most effectively, say for medicinal purposes.

A fundamental polarity exists in the mineral world between the rock-forming minerals such as mica, feldspar, and quartz and ore-forming minerals such as iron, pyrite, stibnite, and galena. The rock minerals and the ores in a hot fluid state only separated from each other in the veins. They can also be rearranged and finely distributed in sedimentary rocks. Suppose one asks about the formation of the basaltic and granitic massifs from which both mineral groups originated. It is clear that the basaltic formed the solid planets and the cosmic bodies, although it only occurs on the Earth under the oceans and continents as a solid shell. Everything that follows above this earthly basaltic shell is within the known cosmos unique to the Earth, including the granite, gneiss, and sedimentary layers of the continents with their vast amounts of graphite, coal, oil, and natural gas deposits, which were formed from the remains of living organisms. Nor can the geosciences explain where the 360 billion km^3 of ocean water, the 10-12 km high air layer, and the living natural kingdoms originate. Suppose the general structure of the Earth with its variety of phenomena is observed in detail. In that case, it becomes clear how every development in the Earth's history, right down to the mineral level, began with a uniform whole that was increasingly subdivided. This assumption leads to the idea of an original earthly life sphere. The minerals above the basaltic, the water and air spheres, and the living environments separated and formed the Earth's nascent surface.

What If one adds Rudolf Steiner's spiritual scientific knowledge to this. In that case, a comprehensive picture of the mutual evolution of Earth and humankind can be gained, which makes it possible to understand the continuous separation of the mineral substances, the plant and animal worlds from the development of human beings. About the connection between the evolution of the Earth and the human being, between the world and the human being, Rudolf Steiner says that everything outside of human nature was in the course of human evolution set apart, removed from the human. "[...] The human being has had to pass through stages of development only possible because certain essential processes take their course in the outer world he is faced with, and he is thus enabled to take certain other processes into himself for his use. So that there is always a complementary polarity and kinship between certain external and certain internal processes." [11] Elsewhere it goes on to say: "Where is the beginning condition? [...] You must again go back to the human being, [...] that is, you must seek the beginning state of nature in the physical and etheric organization. It is not only necessary to apply physical and chemical methods to the human being, but also to find, through spiritual-scientific methods of research, in addition to this corpse-like nature, the beginning state that is only contained in the human being itself, and which were once externally real in older epochs of world development, the evolution of the Earth." [12]

It is necessary to reach from the sequence of phenomena to the imaginative recognition of the essence: "One encounters what has become and, in turn, rises in the reverse direction toward the becoming. Forming takes place in a descending direction. One ascends again in the reverse direction and goes towards becoming. So, what one develops as perception and forces of knowledge in imagination, inspiration and intuition always have a counter-effect in the powers of creation, which are expressed in the formative forces, the processes of separation and consolidation." [13]

The spiritual scientific approach is shown schematically in Fig. 2. [14] "The left side shows the descent of the rock-forming and ore-forming substances from out of the starry realm into the life-sphere through various conditions, through hardening and crystallization up to their final condition as minerals of the Earth. On the right-hand side is shown the further working done by humans, including analysis and creating new chemical combinations with particular properties, for example in technology and medical remedies." [15] The connection of humankind with mineral substances is based on the mutual evolutionary process of the etheric body from the spiritual to the etheric state in the warmth-light-air body of the Earth's life sphere during the Earth's Hyperborean epoch. [16] Out of the spiritual world, the rock-forming substances from the realm of the fixed stars and the ore-forming metal forces from the planetary spheres act on the Earth's life sphere, causing them to contract and become concentrated. These two groups of forces still have a subtle effect on nature today. The human germ lived at that time in the outermost periphery of this sphere. [17] In the Early Lemurian epoch, the mineral substances of the granitic masses began to precipitate from the life sphere above the basaltic that

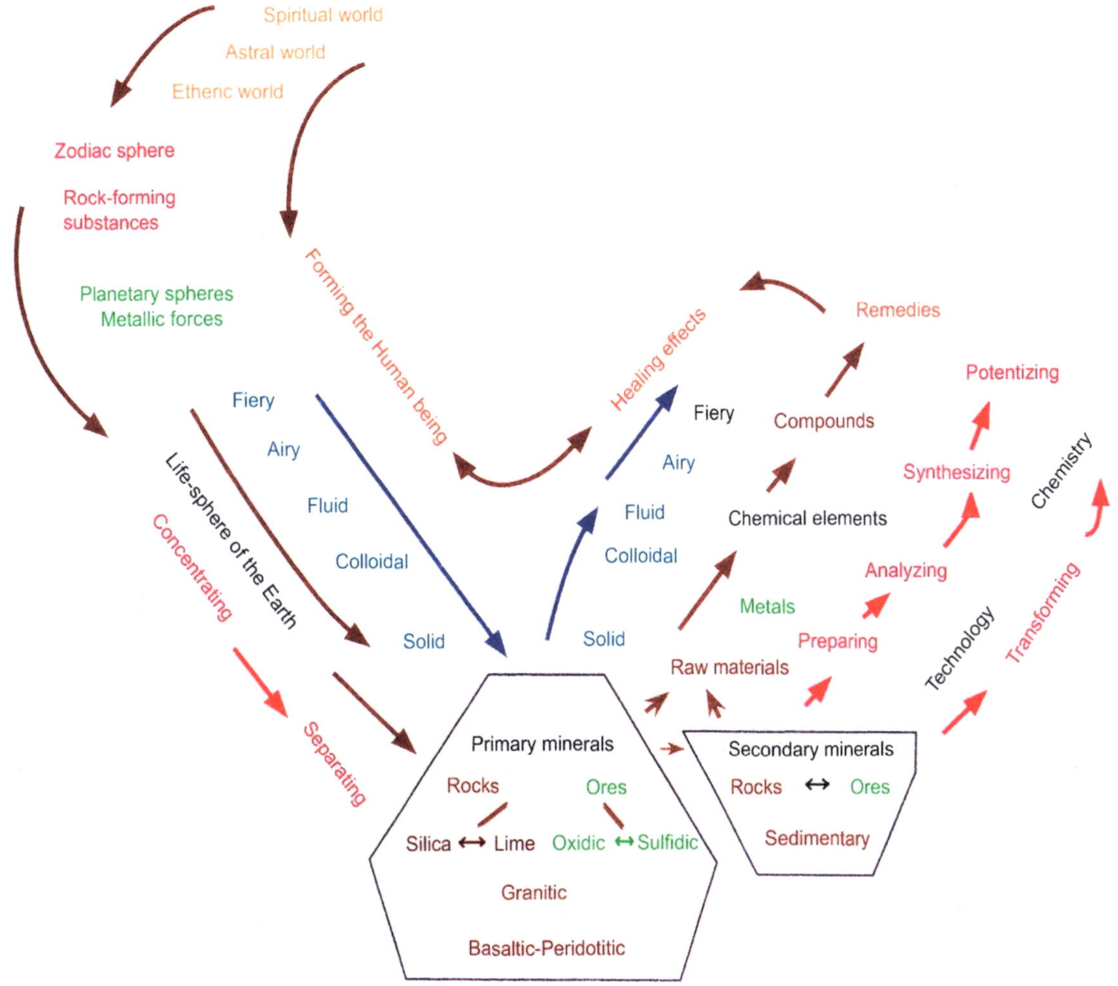

Fig. 2: The descent of rock-forming and ore-forming substances from the life-sphere, through the various conditions of substance down to their end condition in the minerals, and their further utilization by humankind in technology and medicine.

was formed earlier. This activity was true for the continental third of the Earth before it began to rotate. Until the end of the Paleozoic, it was mainly the gneisses with their transitions into the slates that followed and crystallized into the rocks of today's Earth crust, which is about 10 to 20 kilometers thick. Mineral substances penetrated through the cracks and faults of the basaltic and granitic as hot vapors and then finally hardened and crystallized as veins and primary ores in their solid, unique forms. [18] Thus in the processes of evolution, [...] "three polarities can be distinguished. One polarity of the first order is that of the rock-forming and the ore-forming minerals, whose substances are derived from the fixed stars' realm and the planetary spheres. A subordinate polarity of the second-order forms, on the one hand, the siliceous-calcareous minerals with their relation to the plant-like and animal-like, respectively, and on the other hand, the oxidesulfide ores, whose metals are bound to the earthly through oxygen and sulfur, respectively. On the one hand, the third order's polarity consists of the primary minerals that originated from the primary precipitations out of the life-sphere and, on the other hand, the altered secondary minerals. Naturally, there are numerous transitional cases, as in all polarities." [19]

In contrast to the mineral's evolution, the right side of Fig. 2 shows how humanity further transforms and processes minerals for its use. This activity includes the raw materials from rock-forming minerals, building materials such as lime and silica for mortar, clay for porcelain, and silica for glass. In reinforced concrete, the raw materials of the rock-forming and ore-forming minerals are combined, whose characteristics enable maximum compressive and tensile strength with precisely calculable properties. Metals are extracted from ores, especially from the iron ores; other metals are added to enhance the resultant steel and obtain alloys with precisely defined properties.

"To these also belong the mineral medicines. The pharmaceutical utilization proceeds in the reverse direction from the original origin-process, as they lead the mineral and metallic substances by grinding, melting, dissolving, or vaporization, from their solid condition into a fluid and ultimately to an air-warmth condition. Thus the substance vaporized to form a metallic mirror surface approaches a colloidal condition, similar to that of the ore veins before crystallization. Potentization leads further to a condition that is similar to that which prevailed at the beginning of Earth-evolution. Through this, the power latent in the substances becomes effective. It can be used therapeutically, because the human being was the firstborn of the Earth's development and participated from the beginning, from the periphery of the Earth, in the substance processes." [20] These ideas are indicated in Fig. 2 by the curved arrow with two tips representing the reason for the therapeutic effects of high potencies beyond D23. In atomic terms, the original substance would then no longer be contained in the medium in which it had released its etheric potency. Rudolf Steiner puts it like this: "We reach a zero point where the effects of the substance in ponderable amounts are no longer perceptible. If I continue, the results do not simply vanish into nothingness; but the opposite effects are produced and introduced into the surrounding medium. [...] This medium adopts a new configuration." [21]

To understand the origins of the mineralization process, we must also study these processes in the human being. So from the Earth's beginning, the outer symmetrical shape of the human being was created with its bone structure and sense organs, and then the inner organs were formed with their asymmetrical shapes and functions. Thus the principle connection of the healing effects of the rock-forming substances – the calcareous and siliceous – on the sense organs and the skeleton becomes comprehensible. Rudolf Steiner describes it like this: "The human being contains within himself all the kingdoms of nature. Beyond the human realm that he is still part of, he was related in his formative processes, in his developmental stages, to all other natural kingdoms. In a certain sense, he separated all of these natural kingdoms out of himself and then, in certain cases, took back into himself what he had removed. Yes, it is very important that it is a *taking back into himself*." [22] Concerning a corresponding connection of the metallic, of the inner organs and the planetary forces, he remarks, "[...] that the outer worlds, which are otherwise accessible to our senses, are contracted in these organs, and confront us as inner worlds, as it were, just as outer worlds confront us in the planets." [23]

Of the numerous phenomena that could be described concerning the connection between minerals and human beings, here only one for the rock- and one for the ore-forming minerals can be mentioned. Rudolf Steiner characterized an example in the fundamental lectures for physicians. [24] It should be noticed that when he uses concepts and terms from homeopathy or chemistry, he tries to expand them to characterize associated processes. A good example is when he speaks about the forces of calcium carbonate, the carbonaceous, saline, or alkaline, where the formative principles are meant as the forces of the substances that are active today or were active at the time of their formation, not the crystallized substances that the chemist is referring to. The same is true for silicic acid or silicon, which refers to quartz or the siliceous as a general process. Neither silicic acid nor silicon exists in nature, but only as artificially produced substances.

It is mostly the polarities of forces that Rudolf Steiner uses to illustrate and describe his work. About the connection between the rock-forming and the ore-forming forces, he describes an interesting example that resulted in the sketch shown here as Fig. 2. "Out in nature external to the human being, the entire opposition to silicification is contained in all that we would like to call the process of carbonic acid formation. This process of carbonic acid formation is, to a certain extent, the polar opposite of the silicic acid formation." [25] "[...] If we speak on the one hand of the siliceous process as the force which forms us, and of the carbonic acid process as the force that dissolves us — this perpetual tendency to form and to dissolve represents the process of life. [...] The siliceous would affect the upper human being more and the carbonic the lower human being more. And the forces of the alkaline in sodium, potassium, or alkaline salts, are the forces of balance between carbonic salts, alkalis, silicic acid salts, or silicic acid itself." [26]

It is interesting to compare macrocosmic rock formation examples in Fig. 4 and 5 with the blackboard drawing in Fig. 3. In the gneisses, such a process is solidified from the siliceous via the alkaline to the calcareous. This procession shows the developments within the early Lemurian life sphere (Archaean into the Proterozoic), which took place after the granitic substances separated. In the further development of the life sphere, out of the uniform formative forces of mica and feldspar, the plant-like material first

Fig. 3: The correspondence of substance processes in humans to mineral formations in gneiss. A blackboard drawing by Rudolf Steiner, [27] *silicic acid salts (green), carbonic acid salts (red), alkalis in between (yellow).*

Fig. 4: Red gneiss: orthoclase lens in mica-quartz gneiss.

Fig. 5: Calcite (white) in a calc-silicate lens (greenish, brown) in biotite gneiss (dark).

separated in an etheric to fluidic colloidal state, which Rudolf Steiner aptly described as *de-vegetizing*. This surplus of plant matter was deposited as gneiss, in which the mica content increased, and the feldspars decreased compared to granite, while the quartzes were retained. This process can be followed in detail through the minerals of the mica sequence. The gneisses merge upwards into the mica and clay slates, in which large amounts of organic graphite and bituminous coal are embedded. On the other hand, the feldspars developed within the gneisses from Ca- to Na-/K-containing silicates: in Fig. 4, the whitish lenses in the gneiss are plagioclase, the Ca-silicates, and the large reddish lens is orthoclase, the Na-K-silicate. They correspond to the alkalis in Fig. 3: the yellow wavy line. In the same transitional areas from gneiss to slate where the carbon appears, the feldspar increasingly disappears. The calcareous appears as lenses of dolomite marble, a calcium-carbonate substance ($CaCO_3$); in Fig. 3, the red is carbonic acid salts; in Fig. 5, the white is calcite. The calcium is now no longer bound to the silica, but to the carbon, which is elementally compacted in the graphite but is contained evolutionarily as a whole, as the carbonaceous, not as a *compound*. This entire process of calcification from the feldspars points to the animal's separation from human evolution, a *de-animalization* of the life sphere, as Rudolf Steiner calls it. These developments can be read in detail in the mineral phenomena of the feldspar sequences. Just as the carbonaceous is the scaffolding of plants, the calcareous is the scaffolding of animals and humans. Neither the carbonaceous nor the calcareous are contained in the granites. These huge amounts of carbon are completely new impulses in the formation of rocks, which cannot be explained by the typical materialistic rock cycle. They originate from human evolution, where unsuitable substances were eliminated, leading to the de-vegetizing and de-animalizing of the life sphere.

The lens is the characteristic form in which graphite, the feldspars, and the calcareous appear. It is created by the fact that the substance had collected in a colloidal state, separated from the environment as a sphere – a gesture of the astral – and has been pressed together when it sank from the life sphere to the Earth's surface. Calc-silicates often form an outer shell at the border of the surrounding gneiss. This circumstance is shown in Fig. 5, the brownish zone from the garnet mineral grossularite, [28] which belongs to the calc-silicates. This comparison of the formation of rocks, silicic acid, carbonic acid, and the alkaline in human beings with rock formation is not meant figuratively by association. These processes themselves in humans and the Earth can be compared because of their common origin in the early Lemurian life sphere.

The metallic element is also involved in these rock-forming processes. Metals have been treated extensively in anthroposophical literature, particularly from a medical and educational point of view. In most descriptions, the properties of the solid mineral ore or the prepared metal are usually assumed. The usual materialistic models of mineralogy are assumed to be *correct* for the processes before the ores' formation. The benefit-oriented, chemical way of thinking which proceeds from atoms, considers the minerals to be compositions of the chemical elements concerned, for example, stibnite, which is composed of sulfur and antimony, although this – as with all other minerals – is wrong in its conception and can't be produced in the laboratory. This chemical way of thinking makes it difficult to imagine a living Earth organism in which the individual beings are separated from a whole, as would be the case with Rudolf Steiner's depictions of the mutual evolution of the Earth and the human being.

Rudolf Steiner points to both of these effects using the example of antimony: "Indeed we may say that in considering the region from the heart upwards, we must consider the workings of the silicon process in the human being on the one hand and of what is at work from the metals lead, tin, and iron on the other. Whatever of these or related metals is used is associated with carbonic acid processes." [29]

The white circle on the left in Fig. 6 indicates a section through the Earth, and down under the Earth, the metals are confined in a self-contained form. "[...] But in the environment of the Earth, they exist in this finely-dispersed state; there they reveal themselves in continual radiations so that a constant radiation goes out into the cosmos." [30] This can be seen in the radiating crystals of antimony, which is pictured in Fig. 6 on the right, where it shows that antimony follows certain directions of force in its formation, directions of force that are active in the cosmos. And in another place, Rudolf Steiner adds: "[...] We can literally see the extra-earthly forces of crystallization entering the earthly domain. [...] What is this antimonizing force from above the Earth? From the planetary perspective, it is the interaction of Mercury, Venus, and the Moon. When these planets work together rather than separately [...], they work just as antimony works in the Earth. [...] The forces that work out of the Earth from all of its antimony is the same as the force that works down on the Earth from these three planetary bodies above the Earth." [31]

Just as Rudolf Steiner describes the formation of antimony as originating from a constellation of the planets below the Sun (copper, mercury, silver), it is obvious that arsenic has formed from a constellation of the planets above the Sun, containing the forces of lead, tin, and iron, which is mainly crystallized in arsenopyrite. The ores were all formed in the Earth's life sphere together with the granite, in whose colloidal, hot-fluidic mass they were irradiated by the planetary forces in very finely distributed quantities during the Hyperborean epoch. The planets above the Sun were formed from the solar system's contraction during this epoch, and later the planets below the Sun in the early Lemurian period; their forces could have a metal-forming effect on the nascent Earth. Man still lived in the higher realms of the life sphere.

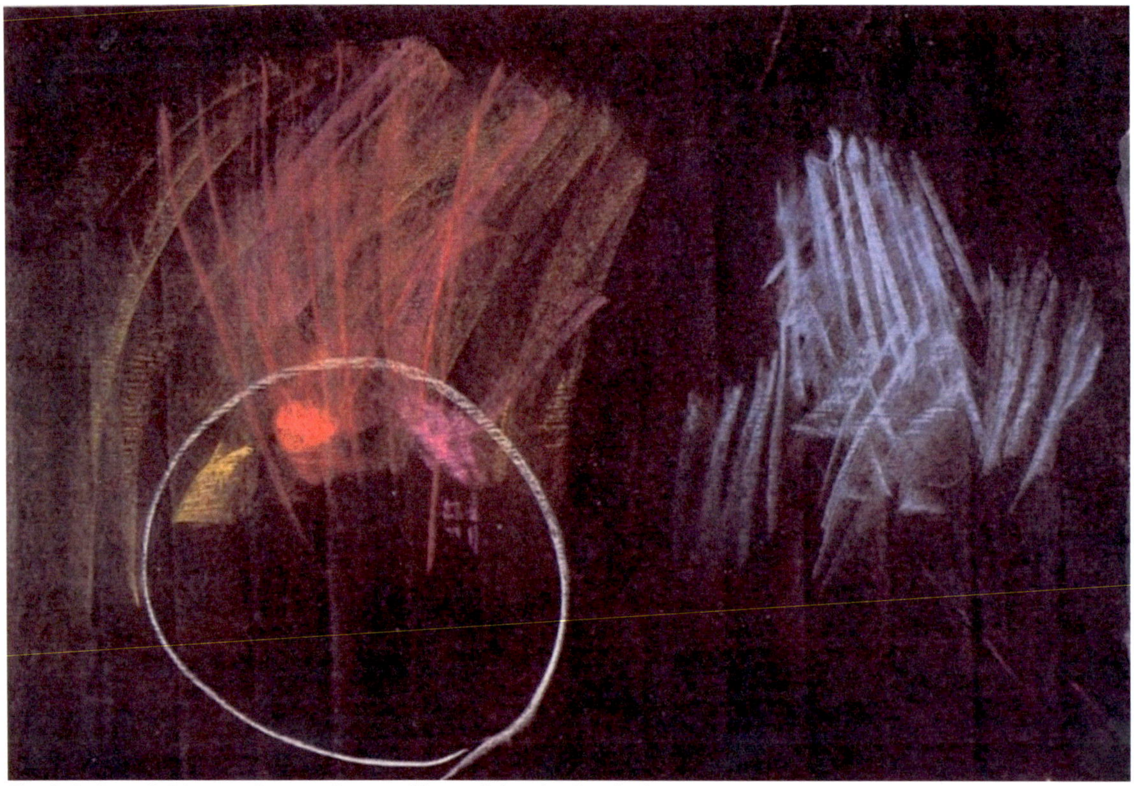

Fig. 6: Steiner, R. Mystery Centers, Lecture IV, panel drawing for the lecture.

Consequently, the life sphere was *de-metalized* – if one applies the formulations of Rudolf Steiner to it. Only after the masses of gneiss had separated and were hardened over the granite did the metallic concentrate within the still colloidal-fluidic granite. And then, in a hot-airy state, the metallic penetrated from below into the fissures of the gneiss and separated from the matrix of rock-forming minerals, and finally crystallized as ore.

Only when we process and melt the ores of stibnite or arsenopyrite can we speak of the metals antimony or arsenic. The long preceding formative process establishes its connection with the human being, which Rudolf Steiner also mentions elsewhere: "In this way, we bring together the metallic quality as defined by the extra earthly element working within the earthly element with what is otherwise rocklike, which takes shape under the influence of the carbonic acid principle and the silicic acid principle." [32] "For example, we must study the human being by tracing the entire antimonizing process [...]." [33]

From the mutual evolution of Earth and humanity depicted by Rudolf Steiner, a connection between the mineral substances and humankind's current life processes can be understood. A new overall picture becomes possible, one which connects the details of the mineral substances, their origin, and the human being's life processes, thus becoming the basis of rational therapy.

Revised text of the first publication in Germany: Bosse D.: Warum können mineralische Substanzen auf den Menschenwirken? Der Merkurstab 2017;70(2):102-109. DOI: https://doi.org/10.14271/DMS-20764-DE

3. Dino-mania. Comment on Steven Spielberg's film "Jurassic Park."

Dinosaurs have been inundating our lives since this movie was made in 1993, exaggerated and multiplied as it was by the advertising industry. You can find them everywhere and in every price range: hundreds of them in toy shops, from videos to inflatable air cushions, from bed linen to shortbread, from bone combs to puzzles, from lollipops to T-shirts, in trade magazines, newspapers, and comics, etc. Over 1,000 products have been licensed in the USA alone with the *Jurassic Park* movie logo, 350 in Germany.

The movie itself cost about 70 million dollars to make. The character John Hammond, the director of "Jurassic Park" says in several scenes in the movie that "we have spared no expense." Advertising expenses were 30 million dollars in the USA, Germany "only" 2 million Deutsche Mark. But this is insignificant compared to the profit of more than one billion dollars earned by advertising licenses to companies; the unlicensed sales cannot be estimated. Besides, at least 250 million dollars was made by cinema operators. On the first weekend alone, 50 million dollars were collected in the USA, more than any previous film had ever achieved. In Berlin, at the beginning of September, there were more than 50 performances a day, or about 250-weekday showings; in the six weeks in which it was already running at that time, there were approximately 1,000 showings. It becomes uncountable if you add to that what came later!

In the beginning, 20,000 copies of Michael Crichton's novel "Dino Park" were sold daily (!) in Germany alone; worldwide, there are some eight million copies. In the novel, it says: "We can make millions from merchandising at the park." It now becomes clear just how far this amount was underestimated. The movie swings from dinosaur park shop items to the products available for purchase in the park's cinema lobby. It is admirable how perfectly the media and marketing interact, from the novel, movie, and advertising (Fig.7) to all industry sectors to natural history museums.

Fig. 7: This is the movie logo for Jurassic Park. The German-language movie advertisement in the Berliner Zeitung of October 7, 1993, is contained with the references. [34]

How did the Jurassic Park movie's tremendous effect come about, which is said to have never been achieved by any movie before it? Paleontologist John Horner, one of movie director Steven Spielberg's advisors, said: "Dinos are like fantasy monsters – only they have lived." There are two important sides to this: the monsters' psychological reality and the physical evidence. You can see the bones if you go to a museum. And then, all of a sudden, otherwise forgotten scientists became sought-after discussion partners! As a geologist and paleontologist, I should give an interview to television from an anthroposophical perspective. But which fragments of conversation from such a quarter-hour interview would then be assembled for a two-minute broadcast? It is most likely that only half-truths would be presented in such a format, which would fit into the common prejudice against anthroposophy while not presenting the whole truths that were intended. But does all of this commotion have anything to do with geological questions?

The bones are the least significant proof of the truth. Advertising would never have had such success if these dragon-like beings were not also occult realities. Manfred Schmidt-Brabant, who was then chairman of the Anthroposophical Society at the Goetheanum in Dornach, Switzerland, describes in his booklet "*Michael Thoughts and Dragon Forces*" [35] how the animality that rises from the lower human being has always been experienced as a dragon. Thus they are suitable images for the present day's soul situation: rapacious with small heads and huge bellies; related to the serpent of the Fall, and only enlarged in a coarse way shortly before extinction. These psychological images are much more effective when they can be proven by the skeletons from the museum.

And what do we know about what is being shown in the movie? How the dinosaurs looked on the outside has not been proven; whether they could roar or in what way and where they lived. The habitat shown presupposes the dinosaur lived in the environmental conditions we experience today, not those present during the Jurassic when the Earth's life sphere provided high ocean water levels that only dropped in the Tertiary. The bones of large dinosaurs are so full of cavities that they could never carry 50 tons of body weight on land; higher water levels present during the Jurassic would make it possible to carry this weight. It is also not always clear what they fed on. The movie is not a scientific journal, although its storyline is about soul matters. The paleontologist is only asked to fill the movie's illusions with the bones' physical reality and thus create psychological realities!

Interestingly, the movie is approved for children and young people from 12 years old, when puberty begins! Where teenagers should have been presented with knowledge about the world, they are instead oriented towards their own body and bound to it! Isn't modern humankind in the same situation where there are no longer any moral rules, where a person can do anything? This path is necessary so a person can make the right decisions out of his or her freedom! In contrast, through today's media such as television and cinema, the spirit is bound to the sensory world.

It is interesting that in 1816 Goethe wrote (Tame Xenia, Part II):

"You can talk a lot about foolish things
Can also write about them too,
It will not kill body or soul,
Everything will remain the same.
But foolishness made visible
Has a magical power because it keeps the senses captivated,
The spirit remains a servant." [36]

The movie dragons of Jurassic Park were born in the Michaelic era: they were discovered in the autumn of 1990, filmed in the autumn of 1992, and released in the autumn of 1993. Some of these dinosaurs consist of steel bones with latex skin and are moved by remote control. However, most of them originated from computers: figures were moved by puppeteers and digitally recorded, rendered as animated models, and then copied at the appropriate size into the scenery. For the first time, this technique led to illusions on a large scale that seemed so real!

These dragons used to be accompanied by a Saint George or the Archangel Michael, who was there to keep the dragon underfoot. The Revelation of John portrays [37] a glimpse of the future, how the fiery red dragon appears to devour the spirit child that is to be born in the soul. And Michael and his angels fought against the dragon. And the dragon fought with his angels. But his power failed, and so there was no longer any place of work for his flock in heaven. The great dragon was cast out, the primordial serpent, both diabolic and satanic in nature, and the deceiver of all humankind. He was cast down to the Earth and all his angels with him.

What is the situation on Earth today? In the novel, the beasts are destroyed. In the movie, they live on and stay on the island with all of their unresolved issues. John's Revelati*on* also describes [38] how the woman is given two wings of a great eagle to flee with the child into the desert. In the movie, it is the helicopter that flees with the children across the ocean. With a few scratches, they escape the charms of the powerful dinosaurs. Unseen are the psychological wounds that fear has torn into the hearts of the watching children. Children as young as 12 years of age may see the film. The most important content is the children's escape from the twenty-foot-high Tyrannosaurus Rex with all its enchantments. One is perhaps accustomed to such horror. Only once does a torn-off arm appear, but otherwise, everything seems to turn out well after all. Afterward, you ask yourself: Why? Who is fighting against the dragons? And you realize that it is always chance that saves the children. And in the background, this question is discussed theoretically in talks about chaos theory. Is chance the present form of the Archangel Michael? It is precisely here at the boundary of knowledge where materialism must stop.

Adults can think about what they have seen. But children are directly affected by the fear that the computer-controlled illusory dragons are spreading. The child's soul has a deep trust in the world. Such illusory images destroy this. They are fascinating and replace living imaginations. Only the lower half of the occult truth is shown – the dragons. Bones prove their existence. The children's souls are drawn too early into the separated spectator consciousness of the adult. The advertising industry produces the dragons that devour the child of the spirit to be born. It is where the real battle of Michael rages.

4. Can the Extinction of Large Animal Groups be Explained by a Meteorite Impact?

A research group headed by L. W. Alvarez [39], the American Nobel Prize winner for physics, published an essay [40] in 1980 in the journal "Science" in which he tried to prove that a global catastrophe caused the extinction of the dinosaurs and many other animal groups at the end of the Cretaceous period. A relatively large amount of iridium was found in a clay layer about 70 cm thick at various places worldwide. It was from a time between the Cretaceous and Tertiary periods, in the Mesozoic and Cenozoic eras, or at the end of Lemuria and the beginning of Atlantis. Normally the element Iridium is rare in sediments and more abundant in meteorites. Furthermore, microscopically fine particles similar to a meteorite were found in this layer with quartz grains broken and deformed by pressure. This is the most important evidence supporting the hypothesis that a giant meteorite or planetary body impacted the Earth.

Such an event is said to have led to mass extinction in the animal world. Finally, there is a cause for the great gaps in animal evolution that have occurred several times in Earth's history and been proven by findings. Alvarez's hypothesis was endorsed by many supporters in the following years, especially by physicists, chemists, and mineralogists. Computer models of an enormous impact were calculated down to the smallest detail, the media and popular science magazines took up the topic, and imaginative images and horror scenes about it became widespread. A trillion-ton cosmic body with a diameter of 10 km impacted the sea with cosmic velocity near the Yucatán Peninsula [41] on the North American continent, where it created a 30 km deep crater in the Earth's mantle. This event destroyed the ozone layer, generated a huge, destructive tidal wave, shock waves, firestorms, and the darkening of the atmosphere by particulate matter. These consequences first triggered an "atomic winter"; then the greenhouse effect caused higher temperatures, acid rain, oxygen deficiency, etc. Together with the extinction of the dinosaurs, this has created a mood of anxiety about such catastrophes to this day.

A particularly clear example of the ammonite group, Figs. 8 and 9 illustrate their evolution during Earth's history, although this also applies to other animal groups. Fig. 8 shows the branching of individual genera and species, as documented from about a thousand individual findings. However, such an exact representation does not prove any connection between the different groups.

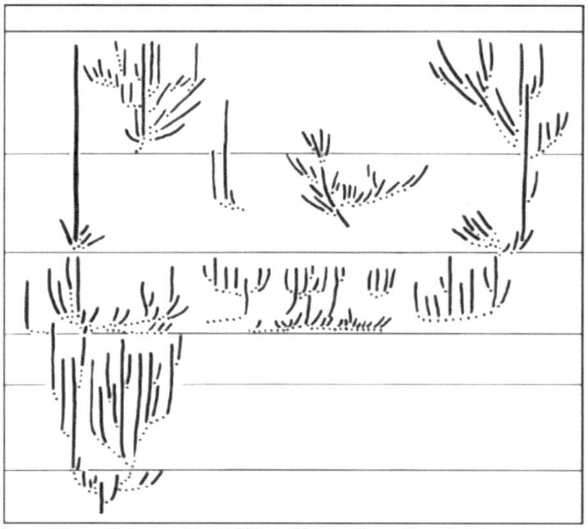

Fig. 8: Branching of the individual genera and species. The appearance in time and relational connection (points) of 186 families (black lines) of ammonites in roughly 1,500 genera, and their range in time. Further details may be found in the text of The Mutual Evolution of Earth and Humanity. [42]

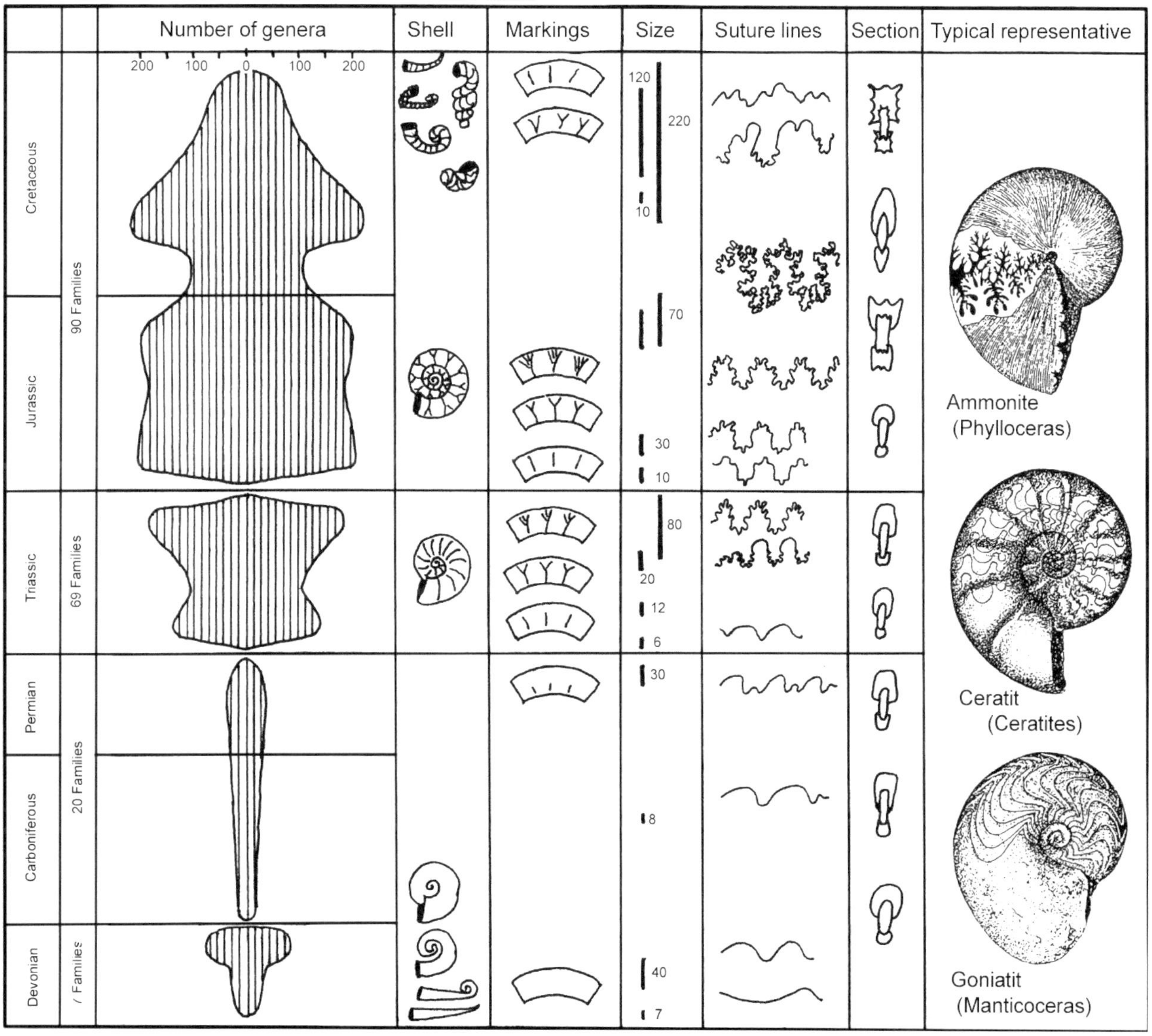

Fig. 9: The evolution of various distinguishing characteristics in time shows an overall evolution in three developmental phases built on one another; far right is a typical representative.

The counterarguments by paleontologists and geologists were more complicated, more difficult to imagine, and not so loud. [43] Some viewpoints will be mentioned here. When looked at more closely, the gaps in animal evolution at the boundaries of geological epochs are far from being as sharp as they appear in broad, popular diagrams. Every species, every genus, follows its inner law, expanding its habitat more or less widely, retreating again, and dying out more or less slowly or rapidly. This dynamic flow of life through Earth's history also has numerous internal rhythms; there are times when blossoming and times when extinction prevails. There are also external influences, such as sea floods, volcanic eruptions, or mountain building. Because of these very diverse conditions, it is not easy to draw "concurrent" boundaries for geological layers worldwide. For this purpose, different habitats of several animal groups separated by great distances must be aligned. The boundaries are then drawn where certain genera no longer occur or where others start anew. Dinosaurs are understandably more conspicuous than foraminifers, microscopic unicellular organisms that are the primary basis for classification since they can be found in almost all sediments. The dinosaurs also show how different the individual genera and groups in the different areas of the Earth are flourishing and dying out, some already in the Jurassic, others in the Lower Cretaceous, and only the very last ones at the border to the Tertiary. A mass extinction did not occur at the boundary between the Mesozoic and Cenozoic, between Lemuria and Atlantis, neither on the land nor in the oceans. One can only find a gradual decline, and after that, many new groups, especially mammals, gradually appear (Fig. 10). Hence, it is justified to establish a primary gap there. It took many decades before it was even possible to agree on boundaries of formations that are valid worldwide.

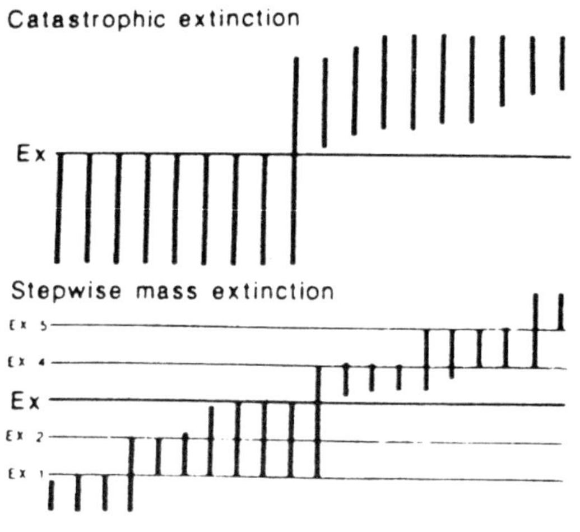

Fig. 10: Two representations of the animals' evolution: Above schematically, below differentiated, as the findings prove.

The other, more important gap in animal evolution is the boundary between the Paleozoic and the Mesozoic, specifically between the Permian and Triassic, the fourth and fifth Lemurian epochs. An Iridium anomaly has not been found anywhere here. Various researchers see global, rhythmic fluctuations in climate as the main factor in the extinction of almost the entire animal world of that time. At the Cretaceous-Tertiary boundary, however, there are no signs of strong climate fluctuations. Such fluctuations must also have had a particularly drastic effect on the plant world. But it was precisely the plant world that continued to develop unaffected by the two most significant "fauna gaps." And as far as the iridium content in the clay layers is concerned, there could also be other causes. Since elevated levels of antimony, arsenic, tin, and aluminum were also found, there is nothing unusual in these layers formed in low-oxygen facies, close to deep fractures, which were probably accompanied by strong volcanism. Central German copper slate contains numerous heavy metals, including, for example, iridium that is three powers of ten higher than at the Cretaceous-Tertiary boundary.

An offshoot of the impact hypothesis has been added in recent decades. Attempts were made to find regularity through time by relating the indicated rhythmic fluctuations of the life stream with millions of years of radiometric data. Nevertheless, radioactive measurements are not possible in sediments, but only where volcanic rocks are embedded. Meteorite impacts were calculated from these measurements, and the cause of solar system oscillations was determined. Such trains of thought seem like a scientific deception.

Nevertheless, Alvarez's impact hypothesis has caused paleontological research to reconsider the basics. It is just amazing how this hypothesis persists. Detlev Thies [44] suspects: "The fascination of the catastrophe still outweighs the well-founded paleontological objections." It is more natural for the physicist to search for external cause-and-effect relationships in the realm of life than to imagine complicated transitional fields of living rhythms that follow internal laws. The whole subject is perhaps more of a psychological-scientific problem because some human beings think more in a physical-mathematical way with definable units. Others think more in a biological way, in terms of fluctuating developments.

Rudolf Steiner mentioned the viewpoints concerning evolution in a lecture as an example of the question of how the boundaries of natural science could be overcome. "Today, this represents more or less the straight line of evolution (Fig. 11, upper horizontal line). But the facts do not show this at all. A real scientific investigation of nature reveals [...] marvels; what scientists say about nature is not true. For, if we consider the facts without prejudice, we obtain [this]." For all groups, "[...] we reach the source of origin of all, where everything is spiritual, and in the further evolution of man we can see that his origin is in the spirit. Gradually man assumed a higher spirituality. The lower beings also have their origin in the spirit, but they have not assumed a higher spirituality. Facts show us this." [45]

The fundamental problem in all discussions concerning evolutionary origins rests on the fact that fossil findings show a discontinuous row of individually bounded types of species, which do not transition into one another but rather lead to extinction. However, there can be read a continuous ascending development up to the human being from these rows. From the lack of transitional forms arises the question: who caused this ascending evolution from the lowest animals up to humans? Within natural science, no question has been so vigorously argued over, as it ultimately has to do with the question: What is the human being? How does the soul-spiritual aspect relate to the physical body?

Rudolf Steiner discussed this regarding evolution and devolution: "We must first dismantle the material so that the soul-spiritual can manifest in us, such that we can experience it. It will be a great moment to develop a rightly understood natural science when it recognizes an opposite factor, which at the appropriate place carries evolution further: where it recognizes the building-up and also the breaking-down, evolution and also devolution. Then it will be understood how the spiritual in the animal and the human – in the human in a self-conscious manner – take hold of matter. The spiritual can grasp the material not because this latter develops up to it, but rather because the material in a reversed process is dismantled. In the dismantling, the spiritual can appear; it reveals itself. Thus we are filled by the spiritual, which is everywhere present, where devolution is present rather than evolution." [46]

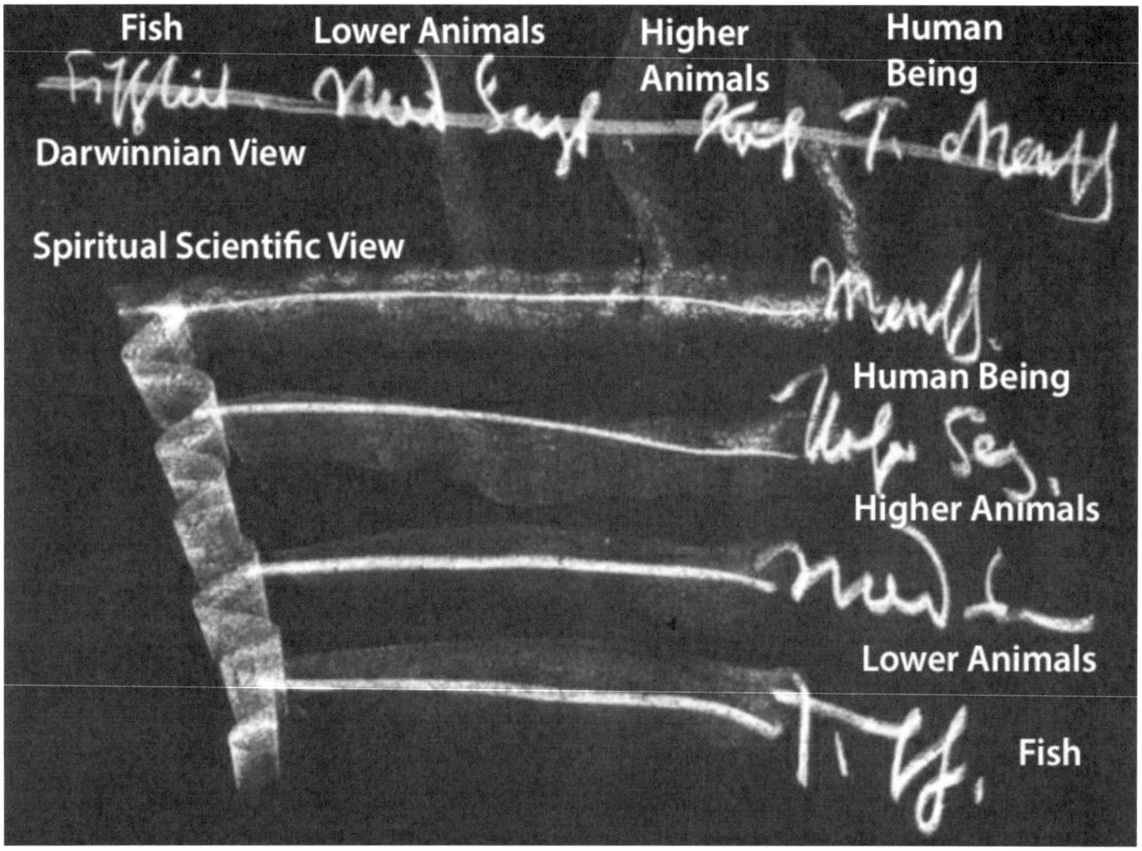

Fig. 11: Ascending evolution from the viewpoint of Darwinism (upper line), as a continuous transformation of animal forms up to the human, and from the viewpoint of spiritual science (below) as a stepwise ascending release of individual groups (right) from a common spiritual origin. [47]

5. The Proof of God in Modern Natural Science: Questions about the unsolved problems of the evolution of a "Creative Universe."

In Berlin on December 6, 2010, in the large public hall Urania, I saw the premiere of an eighty-minute film by Rüdiger Sünner: *The Creative Universe – Natural Science and Spirituality in Dialogue*. Magnificent pictures of rarely seen natural phenomena, which were to be viewed quietly, were placed next to excerpts from interviews that Rüdiger Sünner conducted with quantum physicists, biologists, and the anthroposophical natural scientists Wolfgang Schad, Johannes Wirz, Bernd Rosslenbroich, and Arthur Zajonc. Everything concerned the question of *if and how a bridge between their disciplines and spiritual questions about 'God' and 'Spirit' is possible*.

Most of the pictures provided examples addressing questions about the creation of natural forms, an issue about which the materialistic approach fails. What are the forces that organize the atoms or hold the galaxies together? How could life originate from exploding star clusters or the "primordial soup" which was compared to bubbling volcanic lakes in Iceland? How were the amino acids of Miller's [48] experiment transformed into living, pulsating microorganisms? Who designed the beautiful variety of forms of the unicellular organisms that Ernst Häckel painted? How is the symmetry in the six identical rays of snow crystals possible when every star is designed differently? How is it possible that animals from vastly different groups have developed very similar eyes? Who gives the wonderfully swinging flock of birds its shape? Which artist painted the eyes on the peacock's feathers or the butterfly's wings? Shouldn't there be an *aesthetic sense* in the animals if this is supposed to have advantages in reproduction? How do random mutations cause a higher development, at the end of which the human being appears? It was nice to see how the two monkeys played with the hewn building blocks – but how did they learn to build houses from them over the last *seven million years*? The film showed impressive pictures of such wonders, but the causes remained unanswered!

It became clear that professors of the natural sciences cannot believe in a god. With intelligent contributions, they raised scientific questions: What are *atomic forces*, who provides the *genetic information*, what orders the *morphogenetic fields*, who or what causes the *random mutat*ions, who is the *intelligent designer*? Only the Jesuit priest in the Vatican's astronomical observatory in the Nevada desert declared his belief in God. There I felt taken back 800 years into the dispute between the nominalists and realists and their search for academic proof of God. Great, the film has now brought out the modern boundaries of knowledge! But the film's title promised a dialogue, of which there was none!

In the final interview, Prof. Hans-Peter Dürr confessed that he, as a quantum physicist, had spent 50 years researching the structure of matter and, in the end, concluded that it did not exist at all! That is precisely what Rudolf Steiner said fifty years earlier! Who or what then is responsible for the ordered forms? An apple, he said, has a reality that even nourishes me; but the spirit? Dr. Johannes Wirz was introduced as a representative of the Goetheanum. Besides eurythmy and the visual arts, there is also a Natural Science Section, as the moderator explained, and to which Dr. Wirz even belongs! Nevertheless, he did not mention that the ideal of the Goetheanum is the Anthroposophy of Rudolf Steiner as a path of knowledge! Should it not have been essential to stand by Anthroposophy in the run-up to Rudolf Steiner's 150th birth year in 2011? I was so astonished that I lacked the presence of mind to jump up and say what the Goetheanum is all about!

There were no answers provided by the film, which is a good thing for now. You have to find them yourself. But how? Wouldn't it have been possible and necessary to make mention of the individual path of knowledge? And that Anthroposophy is a path of knowledge that starts with its research precisely at the clearly defined boundaries of knowledge and continues in such a way that a complete understanding of natural and spiritual science becomes possible! Instead, Dr. Wirz spoke – in connection with Prof. Dürr – about the fact that he, too, as a geneticist, was in search of ever smaller units of the living being.

Rudolf Steiner also studied the natural sciences in Vienna in the autumn of 1878 and 1879. And in the same autumn, Pope Leo XIII, in his encyclical Äterna Patris, referred back to Thomas Aquinas and reiterated that knowledge and faith should be strictly separated! Shortly before, Ernst Häckel published his book in which he energetically justified Darwinism. Darwin had made his way in the world from 1837 on. He had seen the giant turtles on the Galapagos Islands, where a devolution toward an end is foreseeable. In 1842, in England, he discovered the thesis of Maltus about the struggle for existence, which he had experienced among English industrial workers. Darwin applied this concept as the driving force in the evolution of the animal world. And since then, this became the real question of materialistic Darwinism, which is still being discussed in thousands of publications and debated in just as many variations! After all, there is no other way to justify evolution if one excludes the spiritual causes!

In the end, a huge question mark remained standing in the room: How can the universe be creative? How did the human being who tries to recognize the wonders of the world come into existence while profoundly separated from them? Apart from faith, is there no "bridge-building" possible to spiritual questions, God, and the Spirit? Or does Anthroposophy perhaps make it possible to substantiate an exact and detailed knowledge of Earth's common evolution and humanity based on natural and spiritual science?

6. The Task of Materialism and its Resolution in the Michael Age

Materialism looks at the world from the perspective that matter is the only reality and the spiritual only its properties and effects. Based on this view, the development of nature and humankind is caused and brought about by purely material forces from which, in the end, the soul-spiritual of the human being has also emerged. Rudolf Steiner characterizes materialism in the last chapter of his *Philosophy of Freedom* as a view with two different sets of facts before it: the material world and the thoughts about it [49]. Materialism seeks to comprehend thoughts "[...] by regarding them as purely material processes. The materialist believes that thinking occurs in the brain, much like digestion occurs in the animal organs. Just as they attribute mechanical and organic effects to matter, they also credit matter in certain circumstances with the capacity to think. They overlook that in doing so, they are merely shifting the problem from one place to another. They ascribe the power of thinking to matter instead of to themselves." [50]

What is the role of materialism in the development of human consciousness? How does it affect the thinking of natural science today? How can its negative effects on people be resolved? How has it been able to develop at all since the beginning of modern times? I would like to characterize these questions using Rudolf Steiner's own words. "Materialism did not originate because material beings and processes alone can be perceived in external nature, but because the human being had to pass through a developmental stage which led to a consciousness that at first was only capable of seeing material manifestations. [...] Only when the thoughts were imprinted in the physical body and when the consciousness extended only to this imprint — only then did the possibility of freedom arise. This condition came with the fifteenth century A.D." [51] "This was brought about by spiritual beings who worked behind the scenes of historical development. Rudolf Steiner gives the times and the names of archangels, who take turns in reigning over humanity's development. Thus from 1510 to 1879, the archangel Gabriel reigned." [52] He was succeeded in November 1879 by the archangel Michael, who will reign until about the year 2400. Natural science developed during the reign of Gabriel. It is "the tendency of this scientific inspiration to create a science, a view that promotes understanding of the material world alone, that is connected with the physical brain." [53] During the age of Gabriel, something was worked into the human physical body from the supersensible world. "During that period, delicate structures arose within the front part of the brain and were gradually implanted into human procreation." [54]

"Towards the end of the age of Gabriel, there took place a battle in the spiritual world [...]. The battle fought by Michael and his followers against certain Ahrimanic powers, [...]. It concluded in the autumn of 1879 when Michael and his followers won a victory [...]." [55] "The natural science of today does not originate from the inspiration of Michael, but one of his companions, Gabriel. This scientific inspiration tends to create a science, a world-picture that promotes understanding of the material world alone and is connected with the physical brain. Within the last few decades, Michael has taken the place of this inspirer of science, and in the next few centuries, he will give to the world something that in a spiritual sense will be equally important — immeasurably more important than the physical science which has advanced from stage to stage since the sixteenth century. [...] So will Michael in the future endow humankind with spiritual knowledge." [56] "From November 1879 on, occult life changed fundamentally. Previously, all occult life took place very quietly, hidden from the outer world under Gabriel's direction. Materialism developed in outer life. It is not something evil, but it has its task in the world plan. Since then, it may be freely carried out into the world under the leadership of Michael. Simultaneously, as a dark counterforce, the influence of the 'God Mammon' began. Not only the god of money but the leader of all dark and lower forces that work their way into the physical body of the human being." [57] Nevertheless, "[...] this materialistic phase of natural scientific development is not underestimated by Anthroposophy. Anthroposophy must see in this materialistic phase of natural science an important transitional stage, an educational method through which one at first learns to submit oneself to the pure experiential knowledge of the senses." [58]

Natural science has described an infinite number of phenomena. "These facts must now once more be permeated with spirit. But it is good — if I may say so without sounding too paradoxical — it is good that materialism arose, for otherwise, humanity would have fallen into a state of nebulosity. Humanity would gradually have lost all connection with Earth existence. When in the fifteenth century the age of materialism dawned, humanity was in great danger of falling victim to Luciferic influences, and of becoming by degrees more and more hollowed out." [59] These Luciferic influences were followed by Ahrimanic hordes, which led to what is "[...] a personally colored Ahrimanic materialistic world view. Materialism had, of course, reached its peak in the eighteen forties, but in those days, its impulses were more instinctive in the human being. [...] From the autumn of 1879 onwards, these Ahrimanic impulses – namely powers of knowledge and of will – became the personal property of human beings." [60] This circumstance had the effect that only the sense-perceptible side of nature was considered. "Materialism was brought into the natural sciences, and naturalists like Ernst Haeckel accepted it unconsciously. [...] Through Haeckel's materialistic bold way of thinking, Darwinism received its present materialistic tendency. [...] It is important that the sense-perceptible was shown first according to its inner context. Haeckel achieved this in a big and comprehensive way. It so happened that those who want to can also see how the spiritual is active in the formations where seemingly only matter is active. [...] The person who knows how to add the spirit to that which the materialist says, studies in this Haeckelism the best basic theosophy. Haeckel's research results establish, so to speak, the first chapter of theosophy or spiritual science. [...] In the same way as the sense-perceptible natural scientific method of research follows the physical-bodily organization back to distant, uncertain primeval times, the spiritual-scientific way of thinking does it relating to soul and spirit. The latter does not conflict in the least with known scientific facts, only with the materialistic interpretation of those facts." [61] "Spiritual science does not speak of the sense-perceptible material world in the same way as ordinary physics or ordinary chemistry, but it speaks of the physical sense-perceptible world in such a way that in all its details it can flow into what it says about the spiritual world." [62]

"Michael sees how human beings observe nature and then seeks to form a picture of the human being from what they call the *Laws of Nature*. He sees them forming their conceptions: "This animal quality becomes more perfect, that system of organs more harmonious, and man arises!' But before the spirit-gaze of Michael, the *human being* does not arise. For in effect, what is thus thought of as being harmonized and perfected is present only in thought. No one can see it evolving in reality, for nowhere does this happen in fact." [63] This way of looking at the purely sense-perceptible has led to the fact that the human being "in this time, gradually succumbed to complete belief in the material world. As a result, human beings had become so paralyzed in their soul-life that in respect of the deepest treasures of the soul, truthfulness had gone. An era which made the world arise out of the Kant-Laplace primeval nebula which densifies into a globe, and in this process engenders living beings and finally humans — could but say: Ultimately such activity must disappear into universal death by warmth, but that will also be the death of everything the human being has developed as moral ideals." [64] If a spiritual view were "incorporated into our upbringing of children, then human beings would grow up who were not [...] developed from a primeval nebula, or welded-together of matter void of morals or being [...] but from that which springs up in their souls as moral Ideals and then would again be one with what stood at the starting point of our cosmic evolution in its purely natural existence. We human beings would then realize that we are called upon to incorporate into the life of nature what we experience as moral ideals. If only children could grow up today under the influence of such a perception, they would be able to take their place in the world in such a way that they would feel themselves as part of the Cosmos and would thereby have a feeling for life drawn from these very forces which they would absorb into themselves with their knowledge of the Cosmos. [...] Anthroposophy will provide a spiritual natural science, a natural science that enlivens the human being, and that which trickles down as knowledge of the spirit in nature will then transform in the human being [...] into social strength. [...] We can get social impulses in no other way than by taking up spiritual cognition from surrounding nature." [65]

"Between the years 1841 and 1879, a decision was to be taken as to whether a certain store of spiritual wisdom could be made sufficiently mature so it could gradually trickle down to the Earth from the last third of the nineteenth century onwards. Meaning that it comes into human souls and stimulates spiritual knowledge in those souls. Such knowledge can be described today as that of Spiritual Science. [...] Humans must be resolved to look into that realm where materialism sees the *nothingness* and there see [...]

the spiritual world." [66] When this knowledge is stimulated, the human being remains free to follow those inner words or not. "Because Michael has conquered the Ahrimanic powers that came into us, we gain for ourselves a piece of human freedom. [...] [But] there is always the danger of human beings who remain in materialism, in the materialistic, Ahrimanic way of thinking, and carrying this on into ages when, according to the plan of things, it should have been resolved." [67] "But what is necessary is the following: in the future, we must cease to differentiate abstractly between the material and the spiritual, we must instead look for the spiritual in the material itself and describe it as such, and we must recognize in the spiritual the transition into the material and its mode of action in the material. [...] Then one will not have abstract matter and abstract spirit, but spirit, soul, and body working into one another. That will be a Michaelic-culture." [68]

7. Insight into the Goethean Method of Knowing

Rudolf Steiner ends his book *An Outline of Occult Science*[69] with the astonishing certainty that natural science's present and future observations will fully confirm what supersensible perception reveals. With this concluding thought, he takes up what he had already postulated in the preface, namely that "the statements in this book agree with every forward step of modern science."[70]

But let's compare Rudolf Steiner's remarks in the long chapter on The Evolution of the Cosmos and Man with how the Earth's history is described today. We find ourselves confronted with two completely different, indeed opposite worldviews. In geology, the Earth is said to have emerged from a molten state that is cooling down – not from life processes; the human being appears only at the very end of development and not as its starting point; random mechanical and chemical processes would ultimately stand as the cause behind all evolution and not the activity of hierarchical beings.

Nevertheless, there is no gulf between the descriptions of natural science and Rudolf Steiner's research over which a "bridge" would have to be built. There is no need to "prove" Rudolf Steiner's statements through corresponding evidence from the geosciences. Through the Goethean method, we can learn to perceive the spiritual in the physical world's phenomena – just as we can see the human being as a physical being and as living, ensouled, and spiritual.

The **first step** on the Goethean path of knowledge is also the first that must be achieved by natural science in its field of knowledge, namely the thorough, unbiased **observation** of the phenomena, as is done by the geosciences, but to the extent possible without any presupposing ideas, theories, or models. Of course, this cannot be about the thoughtlessly experienced pure *phenomena* because observations cannot exist without concepts. More importantly, *questions* about nature can indeed open and guide the field of observation. Thus a solid basis can be formed to understand the development processes gradually in further steps.

In a **second step**, the **sequence of the phenomena** is important, either in the order given by nature or as the researcher logically constructs it. The time sequence of the phenomena may also have condensed in spatial juxtaposition. For example, in the mineral world, this would be the sequence of superimposed layers; or the sequence of minerals from the surrounding rock to the inside of a druse; for the plant, the sequence of leaves, etc. To read the relationship between the juxtaposition and the succession makes it possible to recognize the processes of becoming. The most important thing is to pay attention to the **transitions**, distinguishing or connecting a phenomenon with the adjacent one. Goethe recommends that the best order is the one whereby various phenomena "become, as it were, one great phenomenon," the parts of which relate to each other. The ability we need when comparing the individual parts is what Goethe calls "precise sensory imagination." To transfer one phenomenon into the next adjacent one, we need imagination, but it remains exactly connected to the perceptions and does not create arbitrary patterns. The point is to carry out in your mind the transitions both forward and backward in an artistic way. Thus our ideas become mobile so we can enter into the processes of becoming by inwardly recreating the sequences of the phenomena and thereby experience the transitions. In this way, similar to a person's life story, we arrive at a biography that can reveal something of its essence to us and lead us further to an overall picture. It requires knowledge suitable for organic life, although, for the description of mineral phenomena, knowledge suitable to the inorganic is applied. Rudolf Steiner describes the difference as follows: "In the process of cognizing the inorganic, concepts are linked together to survey the connection of forces affecting nature. When cognizing the organic, one concept must be allowed to grow out of the other so that, in the progressing, living transformation of concepts, *images* arise that display the being formed in nature. Goethe worked towards this by trying to hold in his mind a pictorial idea of the leaf of a plant – not a static, lifeless concept, but a concept capable of assuming the most diverse forms." [71] Goethe thus arrived at an overall view of plant development with the concept of metamorphosis.

We thereby gain in a **third step** what Goethe calls "pure phenomenon, archetypal or primordial phenomenon," which "is ultimately the result of all experiences and experiments." The human mind defines the variable and excludes the random. This action leads to a vivid overview, a **general concept**, in which the particular case can be placed. Goethe describes this as a "succinct point, a truly spiritual place wherein a whole sequence seems to come together." [72] Much can be derived from there, and further sequences can be connected or related to each other. Thus Goethe had found in the granite of Karlsbad the starting point from which he could deduce further developments of the rocks and minerals. Summing up these first three steps, he said: "I, therefore, had to adhere to my former approach, which compelled me to observe all natural phenomena in a certain sequence of development and to accompany the transitions back and forth attentively. For this alone enabled me to arrive at a lively overview from which a concept is formed that will then encounter the idea in an ascending line." [73]

This idea points to a **fourth step** leading to the **idea** Rudolf Steiner calls a "saturated concept." The idea is able from the other side, from the world of the spirit, to reveal itself in its essence. By this, Goethe does not mean an arbitrary thought, conception, or theory, but a spiritually perceivable reality. He perceived it as an inner experience, which is much more substantial than a model abstracted as an idea from the observations. Goethe describes the ability necessary to ascend from the abstracted concept to the idea as "seeing with the **eye of the spirit**." Thus he formulated that when one observes the transitions from granite to the rocks above, it is as if one has seen with eyes how the former balance between feldspar, quartz, and mica was nullified. Each sequence of minerals "must be observed not with the eyes of the body, but with the eyes of the spirit." [74] Or, in another place, he writes that one must "accustom oneself to seeing with the eyes of the body and the spirit at the same time." [75] To Henrik Steffens, he wrote "that when we look at nature, ideas are awakened to which we attribute an equal or greater certainty than itself. We may let ourselves be guided by these ideas, both when we search and when we organize what we find." [76]

This last already points to a **fifth step** where the discovered idea is **compared** again with the **phenomena**, tested against perception, engaged in dialogue. A more comprehensive "saturated concept" enables us to see the phenomena in a new light, to rearrange them, and add new experiences to them. Goethe writes that when we have gained an idea, it is necessary that we then turn back again to the world of phenomena, which reveals itself in "a thousandfold variety with all variability invariably revealed." [77] He found it very apt when Dr. Heinroth also characterized what Goethe then referred to as follows: "that my thinking is objectively active; [...] that the elements of the objects, the observations, enter into the same [thinking] and are penetrated by it most intimately; that my observation itself is thinking, and my thinking an observation." [78] And Goethe added afterward: "Every new object, well-observed, opens up a new organ in us." By this, he means that the spirit's eye can be formed in thinking itself as an organ of perception for the saturated concept.

As a personal **experience**, it may be expressed how an idea gained can be recognized in this way. The spiritual side of sense phenomena becomes a direct experience, which perhaps has seven characteristic features: First, it is experienced as real as an object of the sense world and as true as a mathematical proof. Secondly, the experience is much fuller and more substantial in content than a mere model or theory. Thirdly, an idea appears as an imaginative image that is conceptually illumined by inspiration. With a little self-knowledge, I notice that I do not conceive it, but it comes to me as if from the outside as if it was given and can be clearly distinguished from my own ideas. Fourthly, an idea can open up a new context, so what was already familiar appears in a new light. Fifthly, despite the experience of truth, even these ideas remain free; they do not become obsessions. Sixthly, an inner weighing up can come about, a spiritual conversation between observation and thinking, phenomenon and idea, particular and general, analysis and synthesis, as Goethe has already described. For me, this transforms both the concept, in that it becomes more saturated and the phenomenon, because I discover new aspects and contexts that I have not previously observed. Something like a personal, friendly relationship develops between the phenomenon and me that involves feeling and will. Seventhly, suppose the idea is presented objectively in this way. In that case, it can also be experienced as evident for other people who have taken a similar path.

In the sense that Goethe intended, most **differences** between researchers can be comprehended when they are only caused by differences in actual sense observation or differences of thought and understanding, but not when the difference is caused by the inability to perceive the spirit within oneself. That is why such differences cannot be cleared up at the level of thought through discussions. It is also not about building a *bridge* between two solid pillars, for example, between natural and spiritual science, but about a continuous path from the outer to the inner observations, which then illumine each other. The limits of knowledge can be gradually exceeded as the forces of thought are trained, meaning that limits of knowledge are individual. In his autobiography, Rudolf Steiner writes: "To me, it made no sense to speak about limits of knowledge; to *know* was to rediscover the spiritual meaning experienced through the soul in the perceived world. Speaking of any limit to knowledge seemed to be a confession that true reality is not experienced spiritually in the soul and, consequently, is unavailable to being rediscovered in the world perceived externally." [79] Thus the path of knowledge presented here can lead to an overall picture that is free of contradictions. "Never are there any grounds for contradictions between spiritual science and true natural science. Contradictions arise only between a sound spiritual science that deals with realities and the theories of dreamers and of those who, while they claim to be standing on the firm ground of science, quickly lose their foothold when their theories do not concur with what the facts proclaim and adhere to what they would like to say about the facts. [...] If you ignore those popular works which declare this or that to be a *scientific fact* and go to the sources, then you will find, especially in the field of geology, that geologists everywhere get to a certain point and then place a question-mark. One can take those question-marks as a starting point for spiritual research." [80]

8. Rudolf Steiner: An Outline of Esoteric Science

Note: The Page numbers in this essay refer to *An Outline of Occult Science*, 1972. [81]

Preface

The Rudolf Steiner Publishing House asked several authors to provide a forward for each of the 12 volumes in its special edition. I was asked to write this essay as a forward to the volume containing *An Outline of Esoteric Science* [82]. I initially leaned away from the task because I felt the intellectual distance between natural science and spiritual science was too great. Increasingly it emerges that bridging an abyss between the two approaches to science is unnecessary because a continuous path between the two already exists, as indicated in the following text.

Characters

Rudolf Steiner concluded the book with the astonishing certainty that present and future results of scientific observation will fully confirm what supersensible cognition reveals, to the extent that sense-perception corresponds to the supersensible event (p. 388). This concluding thought corresponds with the first preface (p. xxvii), where it says that in all details, it can be shown how the presentation of this book corresponds with all the progress made by contemporary science.

But if we compare the thoughts presented in the chapter "The Evolution of the Cosmos and Man" (p. 100) with how Earth's history is viewed today, two completely different, indeed opposite world views emerge. In geology, the Earth is said to have emerged from a cooling molten state and not from life processes; the human being appears only at the very end of this process and not as its beginning; random mechanical and chemical processes would ultimately stand as the cause behind all evolution and not the activity of hierarchical beings.

Goethe could not follow such hypotheses of the world's origin. In his 1785 essay on granite, he wrote: "Enormous masses were skimmed off the body of the Sun and hurled into endless space, thereby creating our solar system. My spirit has no wings to soar into those primordial beginnings. I stand firmly on the granite and ask it to give me access to the thoughts of what the mass from which it arose was like." But how do we ask the granite? It is the deepest rock in the continents' crystalline basement, and in the whole cosmos, it is only present on Earth. The granite was there when the continents were formed and could witness the evolution described in this book. Images of their origin are solidified in the rocks. They must be read. Goethe also mentions this method. In Wilhelm Meister's years of travel, he has the miner Jarno say about the shapes of the rocks: "But if I treated these very cracks and fissures as letters, tried to decipher them, formed them into words and learned to read them completely, would you mind?"

Rudolf Steiner (p. 306) also uses a comparison with the reading of this hidden script to illustrate observations from the world of inspiration. Thus, we can learn to read

Earth's history when we treat its revealed images as letters. We would have to discover these images' natural context and see how they are ordered logically to form them like words through which meaning can eventually shine. In the preface (p. x), Rudolf Steiner writes: "The content of what is spiritually perceived can only be reproduced in pictures (imaginations) through which inspirations speak, which have their origin in spiritual beings intuitively perceived." The congealed forms of the rocks in the sense-perceptible world can be traced back to the processes of their becoming by imagining and turning them into moving pictures that can meet the imaginations. Rudolf Steiner writes: "While reading the communications concerning the spiritual-scientific knowledge, we live in a quite different manner than we do while reading those concerning external facts. If we read communications from the outer sense world, we are reading about them. But if we read communications about supersensible facts in the right way, we are living into the stream of spiritual existence" (p. 18). (On pages 258-260, this path of knowledge is described in more detail.)

In what follows, I want to give some examples of what we can observe through the senses. They are based on many years of geoscientific investigations, which cannot and should not be expounded upon here. Their essence is described in a more pictorial form, just as this book is also written in a pictorially imaginative language. We should not attempt to "prove" Rudolf Steiner's research results through the geosciences or attempt to validate them by today's science. Rudolf Steiner writes, "[...] that the hidden is everywhere demonstrable by its visible effects, [...] and *can* become entirely clear and comprehensible to a person if the manifest processes are placed into the light afforded by spiritual science" (p. 107). A person can find their inner certainty on this path, in the sense that Rudolf Steiner describes spiritual facts (p. 10). Supersensible events can be presented clearly, and thus a personal entrance can be found to the contents of this book. Such individual paths are possible from all areas of life and knowledge.

Granite

Fig. 12: Granite with mica (dark), two feldspars (grey and reddish), and quartz (white).

Fig. 13: Granite porphyry from Goethe's stone collection in the pavilion of the Goethe House in Weimar.

Fig. 14: Granite with a feldspar crystal.

To begin with, through this first series of pictures, let us use Goethe's method to ask the granite "what the mass from which it arose was like." In granite, the mica, feldspar, and quartz can usually be clearly distinguished from one another. The individual mineral grains are closely interlocked because, inside a granite mass, the minerals crystallized almost simultaneously. This result can be seen in Fig. 12. The fluctuation from point to point leads us further to granites in which one or the other of the three minerals predominates. Thus mica can recede in places, and quartz and feldspar predominate. Here the individual minerals are no longer so crowded, and the feldspar begins to form its tabular crystal form, as shown in Figs. 13 and 14. Something emerged from what was initially a unity and then took its course.

In early geological times, during the formation of a mountain range, the still-hot and fluid granite breaks through the rock layers from below upward and pours over the Earth's surface; then, the quartz porphyry is formed (Figs. 15 and 16). As the Earth's already solid surface had cooled down, the mass solidifies and then rapidly crystallizes. Such quartz porphyries were only formed at the end of the Paleozoic period. They penetrated wide fissures in the Earth's crust and are found between older and younger layers in many-hundred-feet-thick blankets spread over many miles.

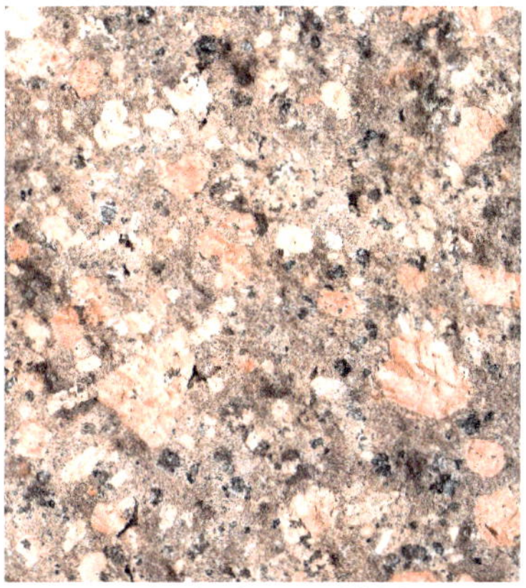

Fig. 15: Individual twined feldspar crystals, which have separated in the quartz porphyry.

Fig. 16: Quartz porphyry with feldspar crystals in fine-grained granitic matrix.

Fig. 17: Quartz crystals from a druse.

Porphyry-like

Other examples lead to rocks where the process of separating goes even further. At the edges of granite formations, feldspar crystals separated and float as larger slabs in an otherwise even-grained granite, as in quartz porphyry (Fig. 15). They accumulated their substance somewhat earlier before the granitic environment became solid. The feldspar substance becomes increasingly pure towards the periphery; its material individuality appears ever clearer; mica and quartz no longer mix in. The crystal form becomes ever more distinct, confining the substance in a flat plane connected with this. This concept can go so far that the crystals separate individually when the surrounding rock is weathered. Fig. 16 shows how their complicated double shape can appear more or less clearly. Quartz could also form its shapes in druse cavities, as shown in Fig. 17. The granitic mass must have been liquid and soft and only then hardened and crystallized to separate in this way. The images of this process were captured at various stages of development. When comparing these individual stages and imagining their transitions, the process's temporal succession can be read from the minerals' spatial juxtaposition. It is impossible to give absolute times because today's Earth has died, and granite can no longer penetrate upward to the surface.

Goethe recognized this process in the porphyry and called it "porphyry-like." He distinguished between the feldspars as "contained" and the surrounding mass as its "container." He wrote to von Leonhard on November 25, 1807, that it is "[...] the entry of the ideal form into reality [...] where it is not possible to see with the eyes, but to some extent, it can be sensed with imagination." The "porphyry-like" process may be regarded as a basic process of all evolution. What is at first united separates materially and thereby takes on form. The individualization in substance and form is thus characterized. Fig. 15 shows this with an example of crystals from a collection site.

We often encounter this process in esoteric science. Rudolf Steiner sheds light on this spiritual process, how it is connected with the cognition process of higher beings: "The whole of evolution depends first upon the severance of independent being from surrounding life; the environment then imprints itself upon this severed being as though by reflection, and then this separated entity develops further independently" (p. 151; also, p. 122). Individualization is always connected with Earth forces. For the human being, it begins at the moment when they take in earthly elements (p. 192). The granites of all continents show us in picture form how the minerals developed within them are varied and individual in substance and form – in all parts of each piece and of all granites on Earth. If we compare the shape of the minerals in Fig. 12, we can find a time sequence. The black mica could clearly define its shape, but the white and reddish feldspars less so, and the grey quartz could only fill the remaining spaces. This time sequence becomes even clearer in Fig. 13. The feldspar crystal shows that it began to crystallize earlier than its surroundings.

The schematic section of the Earth's crust (Fig. 18) shows the crystalline rocks that follow above the granites worldwide: The gneiss and mica schist, which transition into the shales. They transition into the primary sediments of limestones, clay stones, and salt rocks. The sandstones are already secondary from the erosion of all the rocks underneath. Under the granites and the oceans' waters are the basalt and olivine rocks, which only come to the Earth's surface through volcanoes. They belong to the Earth's cosmic development, which it shares with all other solid planets and bodies of the cosmos. Such a picture clearly shows the unique development of the Earth, about which geology describes the visible phenomena while esoteric science points to the supersensible causes.

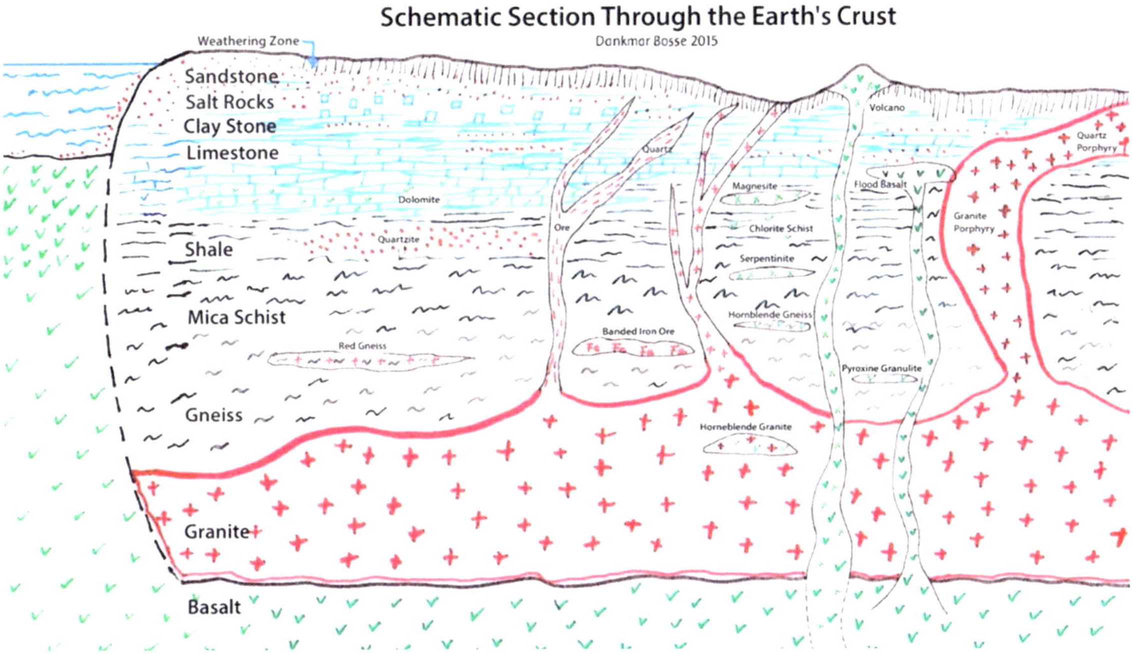

Fig. 18: General rock sequence in the continents and under the ocean

In the gneisses, mica-schists, and phyllites that follow above the granite in the Earth's crust, many other porphyry-like features can be observed, for example, mineral grains of garnet or lenticular bodies of feldspar and quartz and other minerals which also occur in granite (Figs. 19 and 20). New, however, are lenses of iron ore and marble of different sizes. In the white feldspars in granite, calcium is associated with silicon. In marble, the calcium is associated with carbon and oxygen; neither is found in granite.

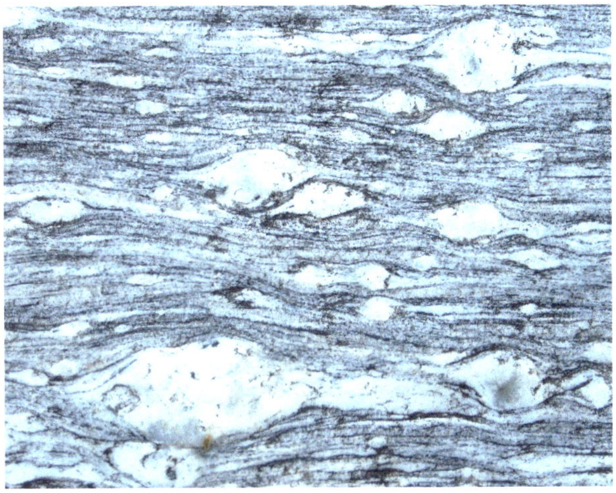

Fig. 19: Gneiss with lenticular feldspar specimens that have separated from a fine-grained quartz-mica mass.

Fig. 20: Gneiss (dark), with a marble lens (top, light) and in between these a shell of calc-silicate (brownish).

The calc-silicates contain both of these elements and also silicon. These lenses must have separated in a fluidic-soft state, similar to the crystals in quartz porphyry. What was a unity previously has separated into a multiplicity. In the fluid state, the substances are separated into spherical forms. The spheres became lenses when the mass was shrinking while hardening. A border around the globule was necessary. For example, on the inside, a substance, a calcareous matter, accumulated from the surrounding area; on the outside, something unseparated remained – for example, the quartz-mica mass. From the white calcium feldspar, a further line of development leads to marble, then further to limestone and animal matter; from the mica, the line of development leads to schistosity, in which graphite appears as a new substance, which then continues further to coal and vegetable matter.

In the beginning, the whole Earth was in an airy-fluidic state and was penetrated by an etheric body; the Earth gradually hardened and finally crystallized. In all that is alive, plants and human beings, all body fluids are in a state in which the ether body can actively be formative and, with death, decay again into mineral substances (pp. 23, 53, 59). This gradual development of the mineral forms is shown in (Figs. 12 and 13), from which one can imagine the transition from liquid to solid. Rudolf Steiner describes this in a lecture as follows: "The further we go back, the more we find that our Earth as a planet is a completely different body, [...] shows us the character of a living being; that we not only find such mineralized air and cloud formations in the vicinity of our Earth as we have now, but we find something within that belongs to our Earth in the most ancient times, like living members of a great living being. [...] On the Earth's surface, there is such a great heat that life cannot develop; but in the environment, life is developing and wants to but cannot descend. [...] And now we look at granite formation and say: originally, the material which the granite contains – quartz, feldspar, and mica – is in a state of dissolution within the great living being "Earth." The latter needs the ability to discard these substances for its development: it separates and lets them fall to the Earth. What is below absorbs what has been separated, builds up a base material, a bone structure in the living being Earth." [83] From this great heat of the early Earth, the mineral substance "burns up," so to speak, and below forms a shell. Below this shell, a very high temperature still prevails today, while above it, water and air which are largely free of minerals have been retained. This situation corresponds to the oldest animal fossils and fish, which only have outer shells, and the horsetail trees of the Carboniferous period, which were hollow inside.

Figures 21 and 22 show notebook pages on which Rudolf Steiner wrote down a few sentences about the common development [84]. Here he uses Goethe's concept of the "porphyry-like" and writes: "In porphyry, the world-plant-animal expires [...] Plant-animal ingrains feeling in porphyry [...] By awakening feeling (porphyry-like) he finds his limbs", namely the human being. In *Esoteric Science*, during Moon development, three kingdoms of nature are mentioned: "plant-minerals," "plant-animals," and "animal-man." They appeared again during the Earth's development as a repetition of the Moon epoch. At the beginning of the Earth epoch, the human being was of astral substance during the Polaris period. Then, parallel to the sense organs, the first soul element was made with the sentient soul's formation (pp. 179, 190). But what does sentient mean? Isn't a boundary necessary whereby a being can feel itself inwardly and feel outwardly something else, something foreign? At this boundary, still higher beings can come to know their ego as a reflection. In this sheath, the sense organs arise between the perceptions outside and an inner world with its desire, joy, lust, and drive. Rudolf Steiner describes the sensation of one's own body that occurs at this stage of development with the words: "That is mine" (p. 181).

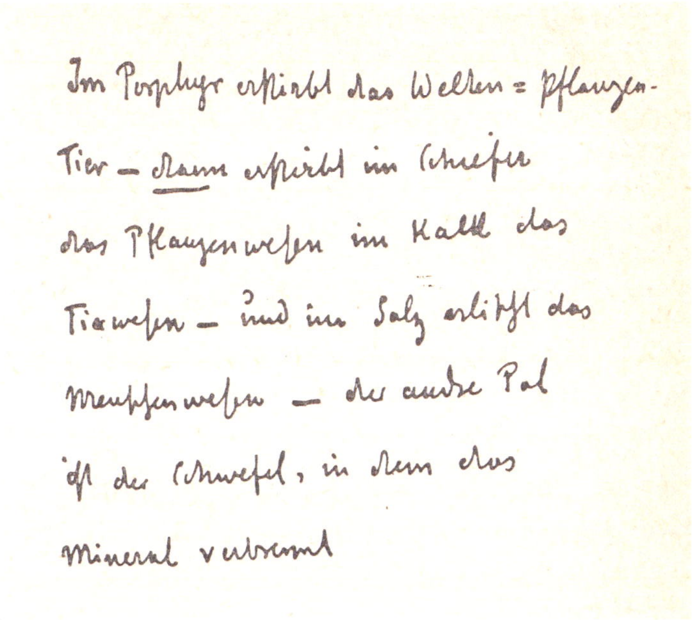
Im Porphyr erstirbt das Welten = Pflanzen-
Tier – dann erstirbt im Schiefer
das Pflanzenwesen im Kalk das
Tierwesen – und im Salz erlischt das
Menschenwesen – der andre Pol
ist der Schwefel, in dem das
Mineral verbrennt

Fig. 21: Notebook page by Rudolf Steiner on the development of Earth and human from 1923.

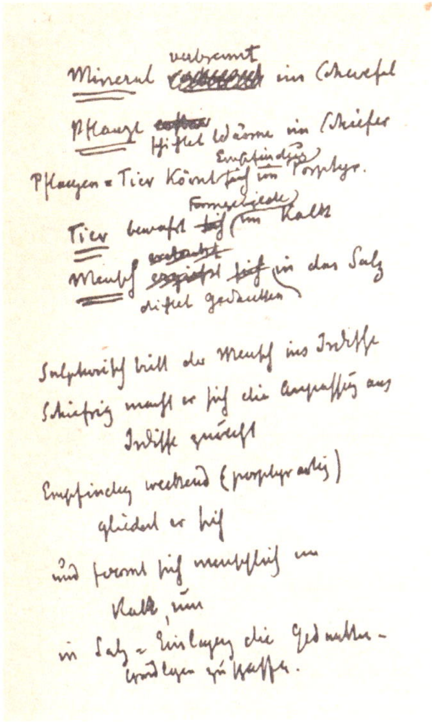

Fig. 22: Notebook page by Rudolf Steiner on the development of Earth and human from 1923.

In porphyry the world-plant-animal expires –
then in the schist the plant nature –
and in limestone the animal nature –
In salt the universal human being is extinguished –
The opposite pole is sulfur in which the mineral is burned and consumed.

* * *

Mineral is consumed in sulfur
Plant spreads warmth in layers in schist
Plant-animal ingrains feeling in porphyry.
Animal conserves forms in limestone
Man condenses thoughts in the salt
With fiery sulfur, man enters the earth
In schistose layering, he makes himself at home on earth
By awakening feeling (porphyry-like),
He finds his limbs and forms himself in limestone
To find the ground for thought in salt depositing.

Fig. 23: Translation of notebook pages shown in Fig. 21 and Fig. 22

The "world-plant-animal" of the Old Moon period expired when the human organized and structured itself porphyry-like and separated from its surroundings. Everything that prematurely died and separated from human development sacrificed itself and, in the hardening process, sank into the solid rock. The spheres and lenses (Figs. 19 and 20) testify to this and show how formative forces continue to work in the same direction: forming shells in the life-bearing colloidal process, collecting substance, and forming it porphyry-like. Even today, we know of something similar. Embryonically, all organs develop "porphyry-like"; they are formed in the same way as the feldspars separated from t h e i r undifferentiated environment. Macrocosmically, this was the same process in the past that microcosmically forms the organs in human beings today, especially the sense organs, which were the first to be formed in this way. The eye turns inwards from the skin and becomes clear, just as the rock crystal in the granite's hollow space is formed from the cloudy, unshaped quartz. While the geological sciences try to understand the Earth's former conditions from the present ones, Rudolf Steiner always looked comparatively at the human being. The human being as a microcosm contains what has happened macrocosmically. It is also true for the history of the Earth, "[...] that in a comprehensive survey of the present, which includes the spiritual, the past evolutionary stages have survived." (p. 109). "Within the Earth, evolution is contained what has evolved during Saturn, Sun and Moon evolutions" (p. 363).

Animal Form

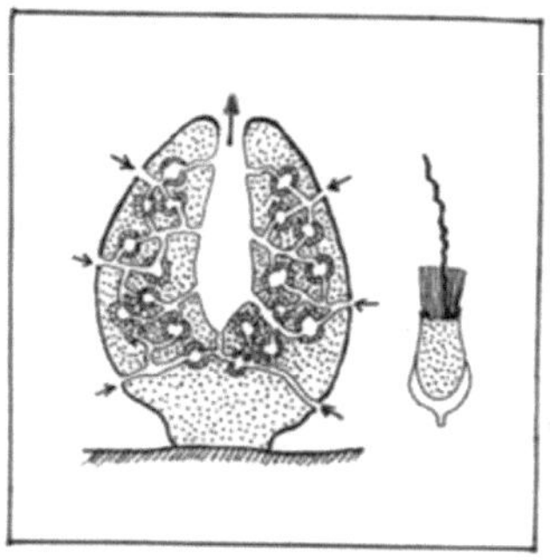

Fig. 24: A sponge with arrows showing the course of the water. On the right is a single choanoflagellate of the type that lines the walls of all the cavities.

Not only minerals and plants but also the animal world separated from humans (pp. 200 and 201). Human forms left behind on Earth became carriers of souls that advanced more slowly and consequently stood at a lower level. So the animals are our brothers! Let us look at the first stage of the evolution of human forms in those Lemurian times. Rudolf Steiner describes them as an "egg-shaped soul form" (p. 180) enclosed by a cup at its lower end like an acorn. The head was already more perfectly formed. The other organs gave the impression of appendages (p. 196). This stage can still be observed very well today in sponges (Fig 24). The flagella move together and create a stream of water that is indicated by the arrows. Special cells also have other life functions which, for example, cause asexual and sexual reproduction. Thus, the sponges stand between the lower unicellular organisms and the higher organisms, and the question may arise: Are they a colony of unicellular organisms, or are they an individual animal? Many related groups exhibit a similar blueprint, even if it varies greatly. The closed head-like pole is directed towards the Earth, the open metabolic pole towards the periphery. For example, in corals, the head pole is surrounded by a calcareous outer skeleton that we find in rock today. They multiplied by growing a new individual within the single mother animal like a branch out of a bud, thus, still asexual. Altogether they formed a "trunk,"; which is an encompassing individual. Suppose one looks at these colored, round, inside radiantly structured beings with movable tentacles. In that case, they appear similar to flower beds and are called "sea flowers."

Rudolf Steiner writes: "[...] at that stage human forms took shape [...] like plant- and flower-like structures of delicate substance [...] that are inwardly mobile, appearing like fluttering flowers" (p. 186). Reproduction and the respiration and nutrition processes also corresponded to the description. The beings that were separated too early at that time in the corals, jellyfish, and similar animals sank and became fossils; their distant

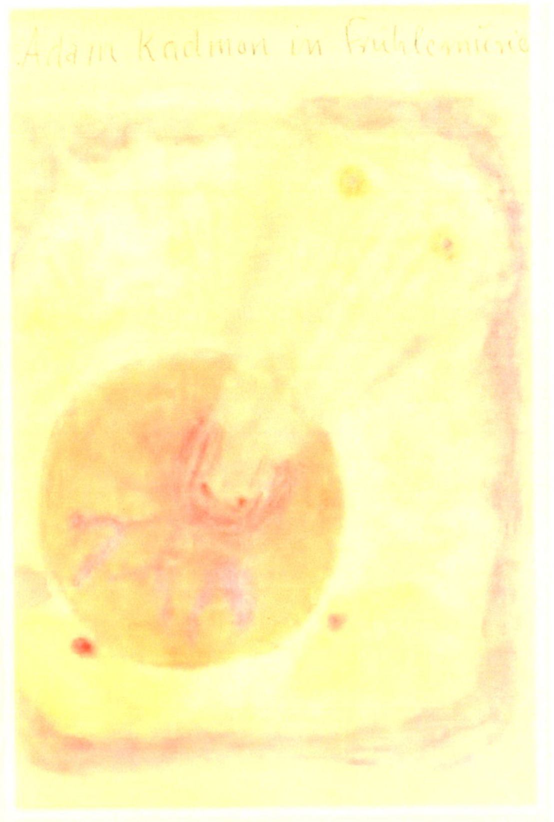

Fig. 25: A pastel drawing by Rudolf Steiner (1923): "Adam Kadmon in early Lemuria. [85]

descendants are still alive today. Figure 26 shows the evolutionary succession of the earliest fossilized animals of the vertebrate group that passes through an embryonic stage when the spinal column has not yet been created. For a better comparison, the drawings have been reduced to a similar size. We first find the early Cambrian echinoderms, whose modern descendants include sea urchins. Then the illustrations show the further developmental stages of a line of animals, and, in the end, at the top appears the sense-perceptible human being.

Additionally, in the embryo, the head is formed first (Fig. 25). The corals' central symmetry now transitions into a five-branched radiant form and then into a bilateral symmetry. The animal formed a self-contained outer skeleton, a shell that is only connected to the surrounding water through small openings. In a lecture, Rudolf Steiner describes: "Around the middle of the Lemurian epoch, [...] another developmental state of the Earth follows. It became more compact, more solid. The human being was thereby forced to adapt to this more solid structure of physical Earth. Human beings could only do this by giving up a part of their being to the coarser materiality. From this part [...], the first imperfect animal world originated. [...] We see these last creatures, which the human being cast off, deposited in the first layers of Earth. These are crustaceans, shellfish which the human being separated from the human form." [86]

A next, astonishing evolutionary step follows. In early Devonian fish, the outer shell is removed from the fish's back and remains only around the head. At the same time, a bony inner skeleton, a spinal column, is built up. Rudolf Steiner describes it as follows: "Man was already present in all that I have shown you, but he was not yet there as a physical being, was there in a very fine body, in which he could be in the air as well as in the viscous liquid." And then it goes on to say: "[...] For when the skeleton of the Earth formed outside, the bones formed inside the higher animals and human beings." [87] The limbs are formed in the opposite direction, from folds of skin, as it were, coming from outside, as if "thrust in" from the cosmic perimeter. Only later, in the case of the

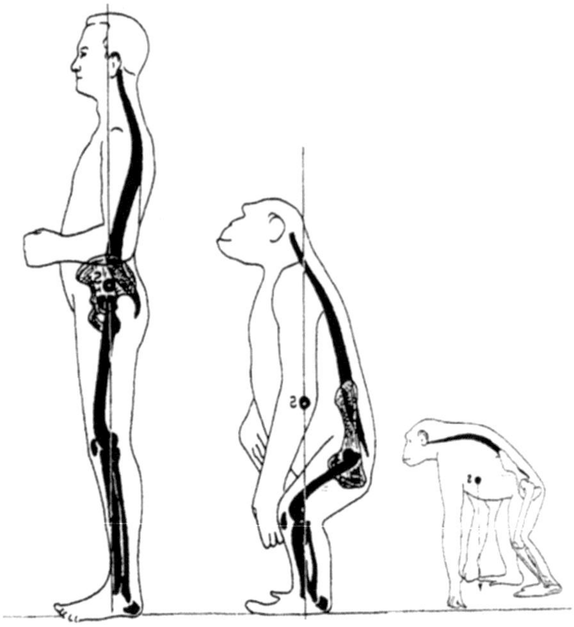

The erect human being, and the monkey, whose normal posture is bound to the earth.

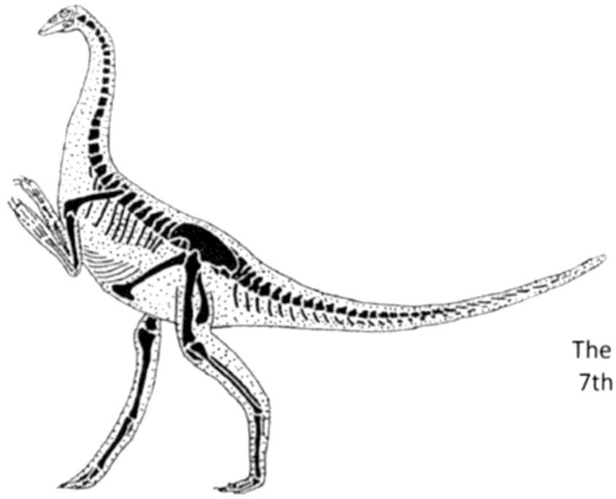

The reptile, a dinosaur from the Cretaceous period, 7th Lemurian epoch.

An amphibian, late 4th Lemurian epoch.

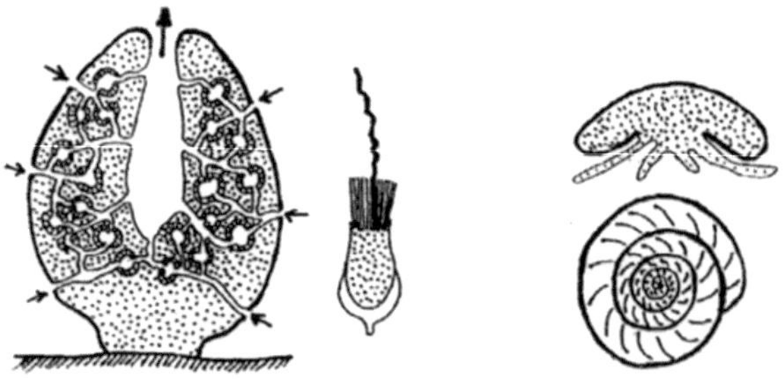

Left is a sponge; in the middle is an individual cell from the inner cavities; on the right are unicellular organisms in section and top view, from the early 4th Lemurian period.

Fig. 26: Some examples of vertebrate development during the Earth's history.

amphibians, do they connect with the spinal column. This structural change then makes it possible for the animal to move freely on solid land. The leg bones of reptiles, birds, and mammals become longer and longer, and the body is lifted off the ground. In the embryo, the head is created first, then an internal skeleton, and finally the limbs.

The upright human being has hands that are free for work in the world. He did not come to Earth with them, where they are specialized and used for locomotion. The head is directed upwards and is formed to perceive and think. The astonishing thing is that the entire higher development of the animals runs as if in a purposeful, completely continuous process through the Earth's history, which corresponds with embryonic development. Since Darwin, science stands at a boundary of knowledge: Who is the carrier of this continuous development? Rudolf Steiner already said about this: "[...] And everywhere, as you will see, spiritual science links up with natural science in such a way that the latter on the pure basis of facts provides ample proof of what spiritual science as such has to offer. Never are there any grounds for contradictions between spiritual science and true natural science. [...] Go to the sources, then you will find, especially in the field of geology, that geologists everywhere get to a certain point — and then put a question mark. Arriving at those question-marks, one can take them as a starting point for spiritual research." [88]

These figures of the animal kingdom that are changing in one direction reveal to us the process of becoming a Lemurian human being. This being is self-restrained and a carrier of this continuity in the Earth's life sphere in physical and spiritual terms. At the same time, what is created externally, is increasingly internalized. Animals become more independent of their environment, and their organs become increasingly differentiated. It is a rhythmic process in which hardening and loosening alternate. Even in gneiss or mica schist, the layers of minerals alternate in rhythmic succession in almost all rocks. The finest layers are often overlaid and penetrated by further rhythms and cycles of a greater extent. These spatial layers represent temporal, cosmic rhythms as the Sun alternates in its effects over the day and year. All life processes run rhythmically between expansion and contraction. During the Earth's history, animals and plants developed into a great wealth of forms, died out again, and then developed anew. Deep gratitude can fill us when we realize the one-sidedness that animals have taken from us humans. They are really our brothers!

Separation of the Moon

In the course of Earth's history, all life on the solid Earth was almost completely extinguished. In the 19th century, these breaks in life development were defined as the boundaries between the Precambrian, Paleozoic, Mesozoic, and Cenozoic eras. The Earth became deserted, and the rocks at its surface hardened into enormous continental plates. The entire Earth threatened to solidify too soon. Rudolf Steiner describes that it became necessary to create a balance between the hardening and dissolving forces so the human being could develop further. "Everything that could contribute to permanent hardening in the solid Earth substance was eliminated." As Fig. 27 shows, "our present Moon separated from the Earth at that time" (p. 189). Rudolf Steiner describes this separation for the workers with simple, vivid comparisons. "You see, when we take this living formation of the Earth, then it came about that one day, one can say, an offspring went out from it into universal space. This event happened in such a way that a small outgrowth developed; it atrophied and finally split off [...] So you can know exactly: the Moon was once inside the Earth! What is orbiting out there as the Moon was inside the Earth and separated from it, went out into universal space. [...] Only the Moon's exit brought death to the Earth and with it the mineral kingdom, which is dead. But this is also the reason why today's plants, today's animals, and the human being in its present form has become possible. [...] On the other hand, the Moon, which has in its surroundings what we have here in the Earth's interior, formed a very thick horny mass on the outside; we look up at it." [89]

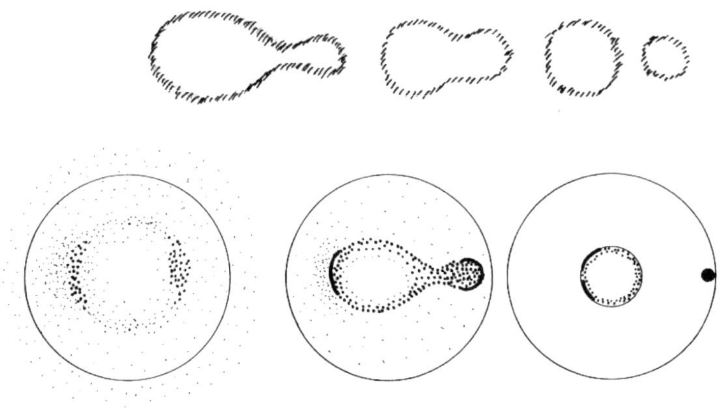

Fig. 27: The Moon's separation: above in a sketch by Rudolf Steiner; below, as it can be understood from the phenomena. There the life sphere is dotted, from which substance was collected at two poles. The black line shows that it condensed to a solid crust and crystallized.

Rock samples were taken during flights to the Moon and reinforce this picture. They are basaltic rocks found under the oceans and in the Earth's upper mantle (Fig. 18). All other solid cosmic bodies are built up in the same way, such as Mars, Venus, and the planetoids between Mars and Jupiter. It is geologically certain today that the Moon once detached itself from the Earth, also by its low gravitational pull. The remaining masses from the intermediate area, which fell onto the Moon in the process, formed the large impacts of the Mare regions and the gravitational maxima among them. They are all on the side facing the Earth, which remains a mystery if one wanted to trace the Mare areas back to meteorite impacts! Most of them were formed in a relatively short time during the Moon formation on the side facing Earth. As meteorite impacts, they would have fallen mainly on the rear side where there are terra areas and very few impact craters to be seen. Today, the Moon only shows one side to the Earth, with which it was once together. The Moon has no water or atmosphere, only completely dead cratered landscapes of basaltic rocks. All present-day hypotheses based on physical causes contradict the phenomenon that a cosmic body hit the Earth and shot the Moon out of the other side, especially since the impact and ejection points could be proven to exist somewhere! By way of volcanoes, the basaltic rocks of which the Moon is composed broke through the Earth's solid crust, which is as wafer-thin as the shell of an egg. And this shell has been torn all along the middle of the ocean and perforated by volcanoes. Today we can directly experience the nature of the Moon's forces through the character of volcanism, which is destructive, unpredictable, irregular, and unrhythmic in connection with earthquakes – a special case in the ordered cosmos. The solidified lava masses still show strong, hardening forces. We can guess what would have happened if these forces had remained connected to humanity and the Earth!

Lemuria

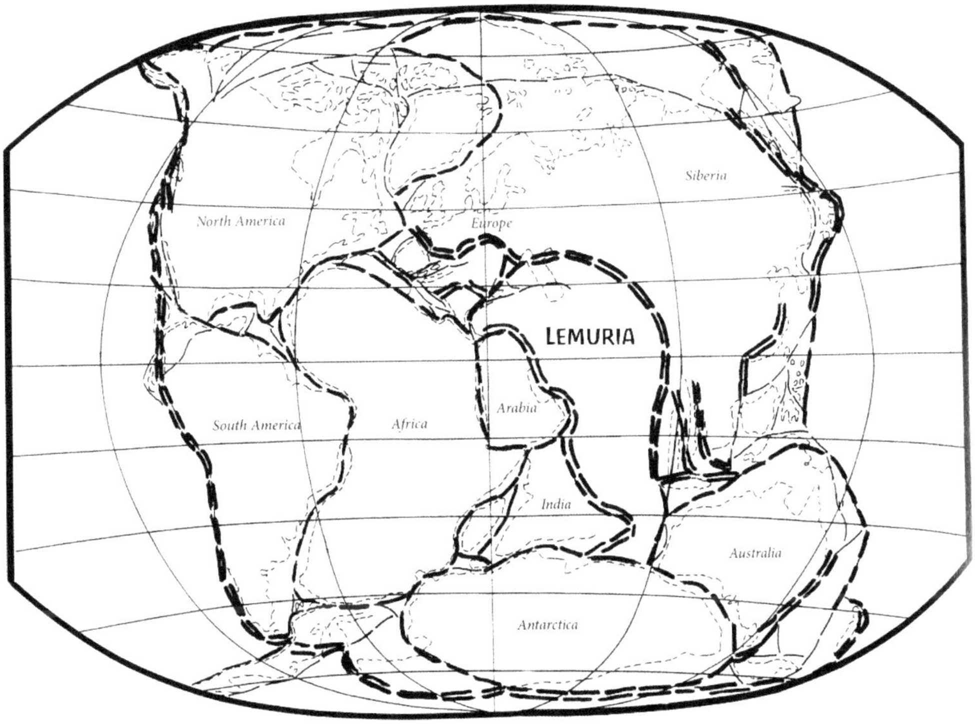

Fig. 28: Location of the Lemurian continent according to today's paleo-geographical evaluations of ocean floor images, between Africa with Arabia, India south of the Himalayas on the one hand, and Central and Southern Asia and Australia on the other.

The Moon's separation from the Earth was likely completed at the end of the fourth Lemurian epoch, between the Paleozoic and Mesozoic eras, possibly in the Carboniferous period. The second great transition between the Earth's Mesozoic and the Cenozoic eras occurred then. Rudolf Steiner spoke clearly of the areas where pre-humanity lived, which as Fig. 28 shows, is "south of present-day Asia, east from Africa, and north of Australia." [90] He did not say it sunk in the Indian Ocean, as maps drawn by Scott-Elliot in Steiner's time showed.

From his supersensible perception, he had before him a picture of the area that, according to the ideas of that time, was not even conceivable at all, and that only today can be geologically and paleo-geographically proven in a more precise way by ocean exploration, plate tectonics, and continental drift. It is distinct from the maps that assume a marine area between Africa and Southeast Asia, representing the so-called "Tethys." But since Chinese geology data is now public, Australia's connection can also be proven palaeontologically, and this also excludes a sea. These processes mainly took place when the Earth and the Moon were still connected. They formed a continental crustal plate on a third of the Earth facing the Sun in the Archean and Paleozoic times, for as long as the Earth was not rotating. On most of the Earth's surface, the hot basaltic masses were seething outside this continental area. The waters of the oceans were still in the life sphere.

On the other hand, the continental rocks bear witness to what Rudolf Steiner described in this way: "The Earth's mineral kingdom also came into existence through having been expelled from the general evolution of humankind. Its structures are what remained and solidified when the Moon separated from the Earth"(p. 203). Above this oldest continental core, the still very soft-bodied human being lived in the Earth's warm-airy-watery life sphere. It goes on to say: "Whereas the human being existed previously like a finely organized being swimming, hovering over the solid Earth surface, he was compelled now to descend from the Earth's environment down to such parts of the Earth as were already more or less solidified" (p. 207). In the life sphere, everything that follows the Moon's development is contained in a germinal state (p. 179) – that is to say, everything that now surrounds the human being in rocks, plants, and animals, the air and water. The life that today has individualized itself in various organisms still permeated the entire Earth. It was a living being, feeling, and thinking. Rudolf Steiner described the Lemurian sphere of origin as "protein-like," somewhat comparable to a hen's egg, but much finer, fluidly vaporous, permeated by a living, differentiated warmth. These substances formed the basis of life for human-animal beings. Granite was the first substance to precipitate from the Earth's life sphere onto the basaltic mantle, forming the future continental core. Today, gneiss, mica schist, phyllite, and slate are layered above the granite. They bear witness to substance processes that accompanied plant nature's death in the Earth's life sphere. According to Rudolf Steiner's notebook (Figs. 21, 22, 23): "Then in the schist, the plant nature expires [...] Plant spreads warmth in schist [...] In schistose layering, the human being makes him- or herself at home on Earth." This cosmic heat is stored in enormous quantities of organic graphite and hard coal. The mineral-plants' comprehensive nature dies into the individuality of the species-specific plant imprints preserved in slate fossils.

The end of the Moon's separation then brought about a thorough break in the living world's overall development between the Paleozoic and Mesozoic eras. The soft-bodied human being separated physically into female and male genders (p. 189). Luciferic beings caused error and illness, leading to death (p. 208). Human beings then descended from the Earth's periphery to the Lemurian continent, to "those parts of the Earth that were already more or less solidified" (p. 207). They became masters of their knowledge, but their ego became steadily dependent on the astral body, which drew the ego deeper into the human form, into the warm, red, iron-rich blood (p. 206). Human beings separated more and more from the perception of the spiritual world and fell into "sin." Through anthroposophy, organs for the perception of the spiritual world can be formed again over time. Because of the Luciferic impulse, the sense world arose in human consciousness and supersensible perception diminished. Thinking awakened at that enveloping boundary where the sentient soul experiences the sense world, which lies between perception and concept and thus awakens the longing for knowledge for the spiritual side of the sense world. Simultaneously, in Central and Eastern Europe, salt rocks many thousands of feet thick were deposited in country-size basins, the largest salt masses in the Earth's history. The human being "condenses thoughts in the salt" [...] to create the basis of thought in salt storage [...] In salt, the universal human being is extinguished," as written in Rudolf Steiner's notebook (Fig. 21). Are these condensed salts the precipitation of the awakening powers of thought connected with the salt processes?

Simultaneously, with the Moon's separation and the separation of the sexes, enormous, as yet unhardened granite masses poured out and solidified very quickly in the cooler air-water sphere near the Earth's surface. There, mighty porphyry layers with billions of feldspar crystals were formed, as shown in Figs. 15 and 16. These are not formed as simple planes, but each crystal is a complex double plane structure, where two individual forms are interlocked as mirror images. Some are hardly separated in shape, up to those which appear to have moved very far to the right or left as if they have been pushed apart, as shown in Fig. 15. The porphyry-like individualization process separated the substances, forming an envelope, and finally caused the tabular crystal shape to separate into the mirrored double form. In the note (Fig. 21), it says: "In porphyry, the world-plant-animal expires [...] Plant-animal ingrains feeling in porphyry [...] By awakening feeling (porphyry-like), he finds his limbs [...]", the human being. One can ask oneself whether these cosmic creative forces also affected the human being's separation of the sexes and separation from the spiritual world.

The further development of the animal world has been imprinted in the fossils of limestones. They make an entirely new appearance in the Earth's history and are embedded initially as lenticular forms in the crystalline shale. They are found in gneiss and mica schist as feldspar lenses and in marble, in the form of compressed spheres of various sizes (Figs. 19 and 20). The oldest crystalline limestones are already clearly permeated by corals, shells, and other lower animals. From primal bacterial life processes, limestone masses several thousand feet thick were formed in the areas of today's high alpine mountain ranges. In the limestone, animal life dies into individual species; it retains forms in the limestone that is an image of the human form at those times. While the plant life spread out in the slates, the animal life turned inward and separated itself. The amphibians with their ossified skulls vividly show the physiognomy of the hardening forces. Then in the Mesozoic era, new animal groups appeared. Reptiles dominated, which, together with the dinosaurs, produced the largest land animals of all. While the amphibians were still poikilothermic, the reptiles first developed their warmth organism; and then they became endothermic like the birds and early mammals. The animals showed a clear tendency to become erect with their bodies, like the dinosaurs or the birds. The air was conquered as their habitat. Reproduction was gradually transferred from the outside to the inside of the body in the early mammals.

At the end of the Mesozoic, in the 7th Lemurian period, the entire animal and plant world again dies out. The Earth also underwent enormous changes. The Lemurian continent, as shown in Fig. 28, separated from Africa and Australia, moved under Asia, and lifted the Himalayas, which are much higher than other high mountains. The American continent drifted westwards and pushed up high mountain ranges. It is the "Downfall of Lemuria." Rudolf Steiner spoke of "catastrophes in Earth evolution that may be called volcanic in nature, where landmasses perished on which pre-humanity lived." [91] The edges of the continents were blanketed by enormous basalt floods, the largest in the area of Lemuria, in southern India, and north of Australia at the bottom of the Pacific Ocean. These flood basalts are shown by green patches in Fig. 29, about which Rudolf Steiner writes: "[...] These human beings could extend the unruly power of their fiery ego's spark to such a degree that they were able to call forth in their environment powerful and destructive fire effects. The consequence was a tremendous terrestrial catastrophe. The firestorms caused a large part of the inhabited Earth of that time to perish and with it the human beings who had lapsed into error" (p. 216). After that, again, an entirely new living environment and stage of human development begin with the Cenozoic period, the period of Atlantis.

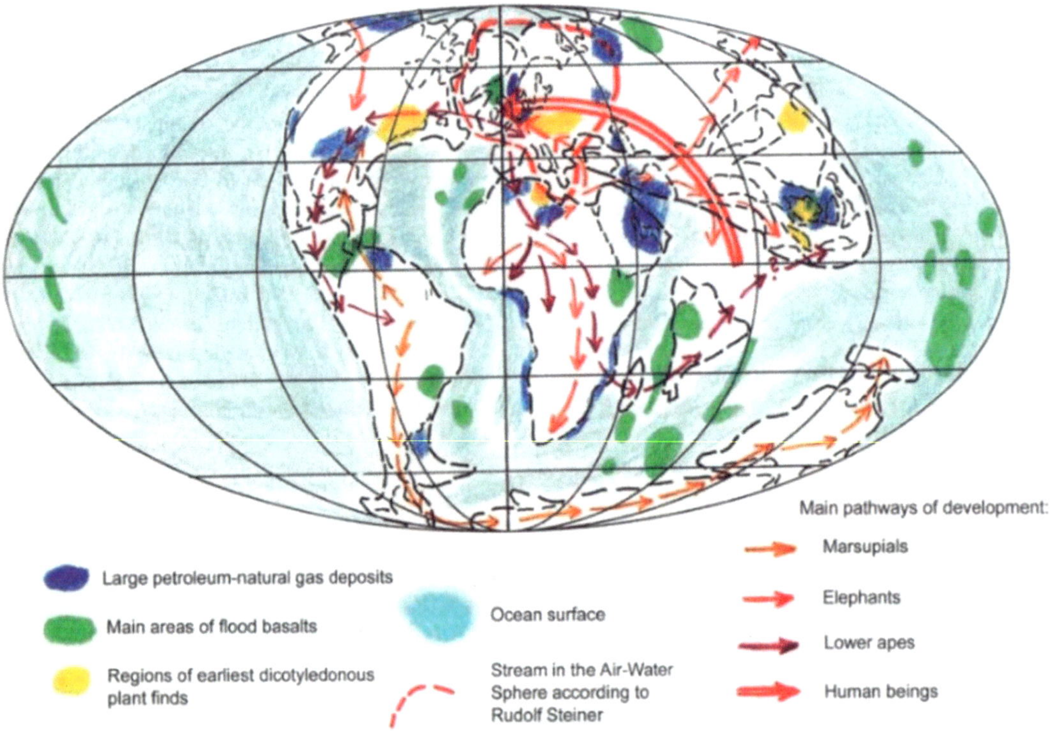

Fig. 29: Migrations of humans from Lemuria to Atlantis and some mammal groups from the Atlantis region.

Atlantis

As with the question about the location of Lemuria, Rudolf Steiner describes the place where a part of humanity was led after the demise of Lemuria. For behind all these developments, the beings of the higher hierarchies were also at work. Only the smallest part, which had remained partly untouched by error, was able to escape to an Earth area that had until then been protected from the corrupting human influence. "Such a dwelling place, which was especially appropriate for the new humankind, appeared in the land that existed on the spot of the Earth now covered by the Atlantic Ocean. It was to this place those human beings withdrew who were mostly untouched by error. Only scattered human groups inhabited other regions of the Earth. The Earth region existing at that time, situated between modern Europe, Africa, and America, is called *Atlantis* by spiritual science" (p. 216). It was a continent located between present-day Africa, Europe, and America. Rudolf Steiner expresses himself in a strangely complicated way, as in the case with the location of the Lemuria, or perhaps very precisely, because he had a picture of a supersensible perception in front of him, which can be understood paleo-geographically only today. Again, he did not take up the usual ideas of his time, as known, for example, from Scott Elliot, who had drawn detailed maps of continental islands "from occult sources." These, and the many similar ideas, have since been thoroughly refuted by geology through oceanic explorations that prove America's horizontal drift to the west. It is interesting how Professor Thenius was able to show that the groups of higher mammals, which developed only at the beginning of the Tertiary from humans who had descended too early, originated from the area shown in Fig. 30!

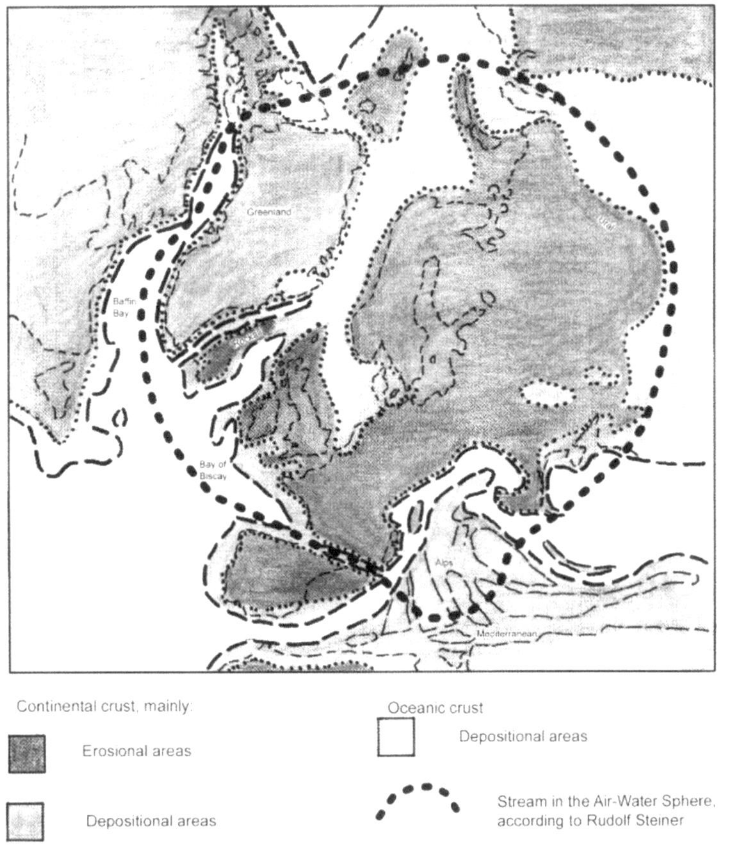

Fig. 30: The area of Northern Atlantic Ocean at the beginning of the Atlantean epoch, during the Paleocene at the beginning of the Tertiary, before the stretching of the ocean floor by the western drift of the American continent. Dark: the erosional areas, grey: the continental depositional areas, white: the oceanic depositional areas, small dots: This is the location of "Atlantis," according to Rudolf Steiner.

At the beginning of the Atlantean period, the air-water life sphere of the Earth's environment had not yet completely separated. A large part of the oceanic water mass was still contained in it. "The human being consisted of a finer, softer substance than the one he acquired later" (p. 222). He did not yet live on the ground, but above it, in the lower part of the air-water sphere. This area was described by Rudolf Steiner in lectures on the folk souls [92] as "a protecting and encircling stream which flowed around the continent of Atlantis" – a sort of precursor to the Gulf Stream. The geographical indications shown in Fig. 29 correspond approximately to the currently known paleogeography of the early Tertiary period, that is, the beginning of Atlantis if we include plate tectonic concepts. At that time, it would still have included Greenland and western Europe and land areas largely covered by shallow seas. The primeval Atlantic was much narrower and had a much lower water level, so it was still separated from the Arctic Ocean; this has been proven by micropaleontology. The mystery centers of which Rudolf Steiner speaks were probably located in the Rockall Plateau area between England and Greenland; today, they are in the narrower area of between 125 and 625 miles, with only one rock protruding from the water. In other areas, where there is now deep sea, sediments were mainly deposited that are comparable to today's shallow-water deposits. Massive salt deposits, such as can only develop in shallow basins, were discovered many thousand feet below the present-day seafloor, like the Gulf of Mexico, the Mediterranean, or the Red Sea. Even in the Pacific Ocean, observations have been made that could indicate a much lower water level.

A new situation arose in the middle of the Atlantean period. Like at the end of Lemuria, the development of the human being led to betrayal and the egotistical use of nature's life forces. Ahrimanic influences were added to the Luciferic influences of Lemuria (p. 226-227). These influences led to the gradual destruction of the Atlantean area by air and water catastrophes on the Earth. "[...] At that time, the Earth received through these storms a new face" (p. 224). The Earth's surface became solid and dry, which the human being needs to live. A firm skeleton was formed (p. 223). The oldest bone finds are from the end of the Tertiary, thus from the middle of Atlantis. The events on Earth responded to this, and today's high mountain ranges were piled up. Only then did most of the water masses precipitate from the atmosphere in tremendous rhythms that filled the ocean basins. The precipitation reshaped the previous uniformly smooth surface of the entire Earth; landscapes were created; deep valleys were carved into the mountains, plains which today are only crossed by comparatively small rivers. Such landscapes are not known from the Earth's history. The water mass filled the ocean basins, whose levels rose, and they consolidated into the Ice Age's glaciers (Fig. 31).

Fig. 31: Atlantis and its mystery centers in the Rockall Plateau area west of Ireland. North of it is another continental fragment towards Iceland. Today, these two areas are 1,640 feet below sea level; they were even higher in Atlantean times. Between Rockall and Greenland, the Mid-Atlantic Ridge – the elongated structure created by the western drift of North America at the beginning of the Atlantean era in the early Tertiary period. Around all the continental borders are submarine canyons.

To such canyons also belong the fjords of Norway, which are carved up to 3,000 feet below today's sea level, or the buried notched valleys, as the Nile valley or the underground channels of Northern Germany. The climate cooled down considerably during the Tertiary; in northern Siberia, for example, there were still palm trees where there are now pine trees, or in the deserts of the Sahara, Namibia, and Arizona, there were still dense forests, which are indicated by the remnant fossilized tree trunks. Seen in this way, the subsidence of Atlantis is a complex event. It was the awakening to thinking, to self-consciousness, to the world of the senses (p. 227); it was the selfish betrayal of the mystery secrets (p. 225); it was the Flood of which the sagas speak that was caused by the egotism of the human being.

The Earthly Human Being

Figure 32 shows a summary of animals' separation from the more highly developed soul-spiritual form of the human being. Group souls separate themselves, which form the main groups of animals in the etheric and subsequently in a soft-body state. Above it, the spiritual hierarchies bring this about through their evolution. Soft-bodied human beings can develop further as long as their bodies remain plastic enough. Individual ego-beings continue to separate themselves from the group ego-beings, creating more specialized forms that develop soft-bodies and appear as new species and as fossils in the sense-perceptible world. Thus evolution goes over into devolution; the soul-spiritual continuity changes into discontinuously separated individuals in a sense-perceptible, physical manner. What in animals is species has in the human being incarnated individually as the ego. Each species retains a certain variability; it goes through a short period of youth, a long period of maturation, a phase of old age, and then extinction. This sequence is highlighted by two examples that are enlarged on the bottom right of Fig. 32. Even the physically embodied human being is subject to this devolution that ends with death. However, their higher self can lead to the next incarnation, which is impossible with an animal species. In the future, in freedom, the human being will develop further and higher in soul and spirit.

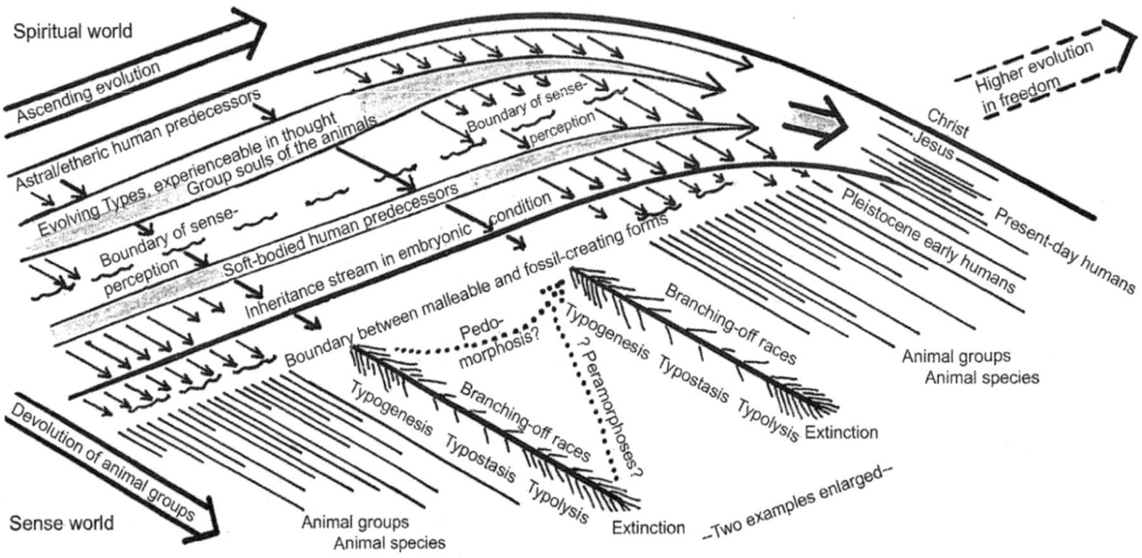

Fig. 32: Diagram of evolutionary directions using the example of the animal kingdom. From the continually ascending macroevolution of the Type, there separate off groups, from which species separate off stepwise and related to one another. They then alter continuously in the succession of generations, which can be termed devolution or microevolution. Their forms harden to the point of creating fossil remains. Given certain conditions, they can further subdivide into races; in the end, their lines end in extinction.

The greatest unsolved problem for materialistic biology since Darwin is: How does a new species emerge when the supersensible side of evolution is denied? Darwin still assumed that it happens by chance through heredity and environment, from the races' final stages, called "peramorphosis." He still lets the very first species come forth from God. The other kind is "pedomorphism," which assumes a random transformation at the embryonic stage. Both are explained today by genes. Still, the question remains open: Who or what programs the genes to demonstrate continuous development throughout fossil history. Biology and modern genetics have already ruled this out.

Altogether, the modern geology of the continents and oceans has revealed an infinite number of phenomena. They still require a completely different world view than the one common today in geology, and such a one was for the first time developed by Rudolf Steiner through his spiritual-scientific research, which is contained in his book *Esoteric Science an Outline,* first published in 1910. In this light, a new overall picture of the human being and Earth will become possible in the future, showing how sense-perceptible phenomena and their spiritual causes could develop (p. 96). Then a mutual understanding between the two will become possible without having to "bridge an abyss"!

Goethe already suspected the connection between Earth and the human being. His friend from Weimar, Johann Daniel Falk, reported on a conversation with him: " [...] in the series of so many different products, through which the creative forces of nature become visible, man is, as it were, the first conversation that nature has with God [...]" And he then put granite in his hand and said "[...] Observe these transitions diligently for me, wherein the end everything in nature ultimately arrives. As you can see, something seeks the other out, penetrates each other, and creates again a third when it is one. Believe me: here is a piece of the oldest document of the human race. But you have to discover the connection yourself. It won't help to tell a person who doesn't find it themself." And so we too closed the arch that started by asking the granite what it can tell us. Its transitions can lead us to the life processes of all the natural kingdoms of the nascent Earth, its rocks, and fossils. All this becomes the oldest document of the human race and its development. This book can help us discover the scientific paths into the connections, but it is still true that everyone must discover and go through it themselves; otherwise, it remains an esoteric science.

9. Experiencing Crystalline Forms

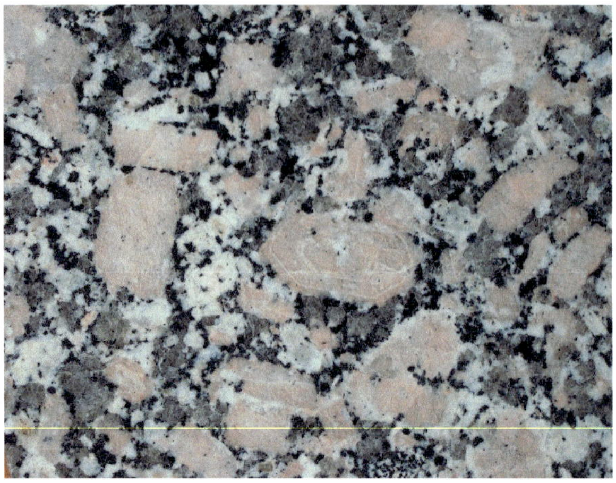

Fig. 33: Feldspar crystals have separated in the granite during its formation.

Fig. 34: A closeup of a feldspar crystal that separated in the granite during its formation.

Fig. 35:A Feldspar twin crystal that weathered out of granite.

Fig. 36: A Feldspar twin crystal that weathered out of granite.

Figures 35 and 36 are from Johann Wolfgang Goethe's collection and collected near Loket (Elnbogen) in the Czech Republic. They were laying in an old field where they had weathered out of granite. They are now in the garden pavilion of Goethe's house at Frauenplan in Weimar, Germany (Prescher catalog 1978, nos. 6889 and 6894).

The feldspar began to crystallize after the mica in the slowly solidifying granite, as a single or double form called "twins" (associatively with human twins). The field of scientific crystallography describes this term as: “Twins are two individual minerals of the same type and formation, lawfully grown together. Twins feign a higher symmetry [...] The substance strives to assume a highly symmetric shape; and it tends to lower indices, therefore coordinate numbers.” The reason for the strictly lawful formation of twins is explained by the random disorder of lattice building blocks. But how can a substance feign something or strive for something? How can such disorder cause the formation of strictly lawful crystals? Who has "put" the two crystals so exactly into each other? Materialistic science is unable to explain this.

In 1807, when looking at such feldspar crystals, Goethe described a different experience from that given by crystallography: "We would like to commemorate the district of Elnbogen, where it can be said that nature has overreached itself with the crystalline form of feldspar. [...] They are double crystals, which seem to consist of two interlocking crystals, but without the possibility of thinking of one or the other individually. [...] In some cases, one has grown on top of the other; in other cases, several are piled up irregularly. Sometimes two such crystals are intimately united in the form of a cross." [93] [...] "No matter how one may describe such pieces, they give the beholder – if not with the eyes, then through sensitivity and imagination – the advantage of allowing the ideal form to enter into reality, which is mysteriously hidden from us in regular, complete, closed forms." [94] Here inner sensitivity allows the eyes of the spirit to perceive how the ideal form enters into reality. In this way, the imagination can use the forms' spatial juxtaposition to read the temporal succession of their becoming.

Goethe showed his collection of feldspar crystals to the mineralogist C. S. Weiss in Karlsbad. Weiss then explained how science names the crystals. They are held in such a way that the flat "top surface" points toward the front. The "normally" oriented crystal is like this. The other one then appears turned by 180° from the right or the left and appears to be "inserted into" the first one so that the flat "top surface" of this part points toward the back. Two bodies cannot penetrate each other in the world of the senses, although it is something that supersensory forms can do. Goethe took his two hands and placed them stretched out on top of each other. Then they displayed a steep "top surface" from the middle finger to the thumb and a flat "roof surface" from the middle finger to the little finger. Weiss was amazed at such a simple way of presenting the forms. Here Goethe had put himself into the process of creating the picture.

Fig. 37: The picture shows models of twinned feldspar crystals and also how the clay was cut.

Figure 37 shows clay models of double feldspar crystals and how the clay was cut out according to the instructions that follow. To create such models, you need clay or a suitable plasticene compound, which should be easily kneadable and cuttable with a wet knife. It is best to use two knives to strip each other of the clay. For this purpose, a container with water is necessary, where the knife is wetted before each cut; otherwise, the clay will stick. In this way, straight and smooth surfaces are obtained. Forming the shapes by hand does not succeed and is contrary to the formation process. Cutting corresponds more to the character of how the surfaces have "penetrated" the substance from the outside during crystallization. First, the clay is kneaded approximately into the form of two rectangular blocks. Since a considerable amount will be cut away, the blocks should be much larger than the finished models. Figure 37 shows the results of the next steps, from back to front.

The surfaces of both blocks are cut straight and parallel (shape 1). Then the steep surfaces on the narrow sides are cut off, which have an angle somewhat smaller than 90°; about 75° (shape 2). Then notches are cut into the narrow sides at the front, back, top and bottom. The upper and lower notches are slightly larger than 90°; about 105°. The front and rear notches are even larger, about 130° (shape 3). Finally, the flat "top areas" are cut to size at the top and bottom according to the right and left shape, respectively. They are extended to the vertical surfaces, cut from the right and left side of each form, and cut out to the vertical front and rear notches (shapes 4 and 5). When the clay has dried a little, after about one or two days, the surfaces can be smoothed or trimmed again with a knife. Have fun and good luck!

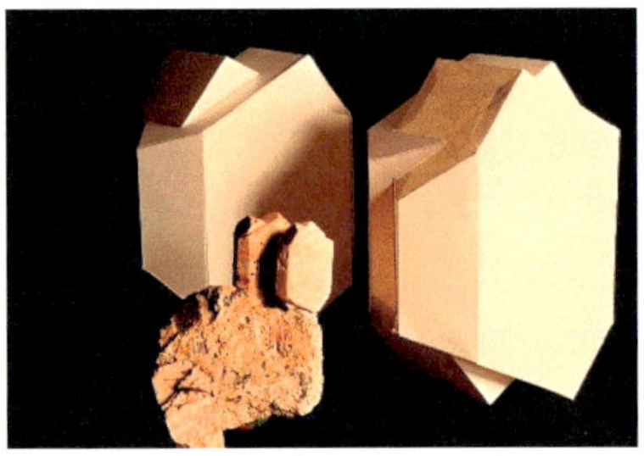

Fig. 38: A piece of quartz porphyry (foreground) and paper models of twinned feldspar crystals.

Fig. 39: A paper model of a twinned feldspar crystal.

Fig. 40: A paper model of a twinned feldspar crystal.

Figures 38, 39, and 40 depict paper models of twinned feldspar crystals. It is advantageous to make the clay models first to gain a clear picture of the paper models before gluing them. A piece of quartz porphyry from my collection is shown in front (Fig. 38), from which the left, rear twinned crystal was prepared; the right, front twinned crystal from the same location is attached to it. Such pieces are very rare. If you enlarge the picture, the shapes can provide a good picture. For the paper models, it is recommended to use firm paper so the final product holds its shape. For the lamp models, you will need sheets of translucent lamp paper of an appropriate size. The cutting patterns are shown in Figs. 44 and 45. The lamps shown are 40 to 50 cm in size. When cutting them out, you should leave a flap for adhesive around all outer edges, which is then folded inwards so that you can glue both shapes together. You have to cut out two pieces from each of the two models simultaneously and work on them further to get a left and a right model of the necessary shape at the end. It is best to place the two finished clay models in front of you, where you can confirm how the forms fit into each other. Finally, all edges should be coated with glue and carefully pushed into each other, which requires patience and skill. You then hold them with your hands and wait until the glue has dried and holds securely.

Fig. 41: Rock Crystal.

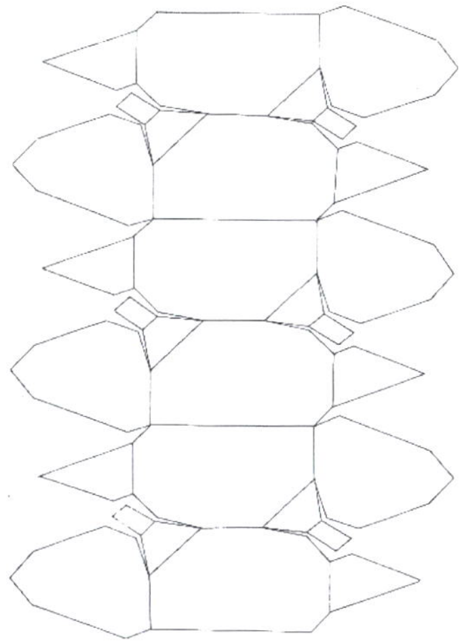

Fig. 42: Template for the Feldspar Lamp Model

Fig. 43: Assembled Feldspar Lamp.

With the two lamps, a larger opening must be cut into the top so the bulb can be changed and for the heat to escape. A bracket must then be attached to the shade and connected to the ceiling with a strong wire. For all of this, good handcraft work is necessary!

Figure 41 shows a rock crystal that has separated from the opaque quartz in the cavity of a druse. It is called a *right* crystal because the right hexagonal tip appears shortened by the vertical prism. Use the same lamp paper in Fig. 42 for this cutout as for the feldspar models above. Since the tips are somewhat angled, it is also recommended to use 40 to 50 cm for the lamp's length, as illustrated in Fig. 43. When cutting it out, you must leave an adhesive flap on all outer edges. This fabrication also corresponds to other feldspar crystals in all other respects. Finally, a large opening for the bulb must be cut, depending on its use. I give additional exercises of this type in the two-volume German-language book entitled *The Evolution of Minerals between the Cosmos and Earth.* [95]

In conclusion, a few words by Rudolf Steiner that correspond to these suggestions for experiencing crystal forms through practice: "In school, our children must learn the different forms in which the various minerals crystalize. They must first learn them according to the laws of geometry, how they can be represented through these laws, and then how they appear in reality, in the mineral kingdom — octahedron, cube, and so on. When we see these forms of the mineral kingdom expressed in geometry, we have before us the original form proper to the minerals. These crystallizations, or rather, these crystal-forms, are in a certain sense the inborn, archetypal characteristic of the mineral kingdom. And the Earth, when this kingdom was incorporated into its cosmic evolution, received at the same time the tendency to crystalize its mineral substances in the forms in which the mineral kingdom crystalizes." [96]

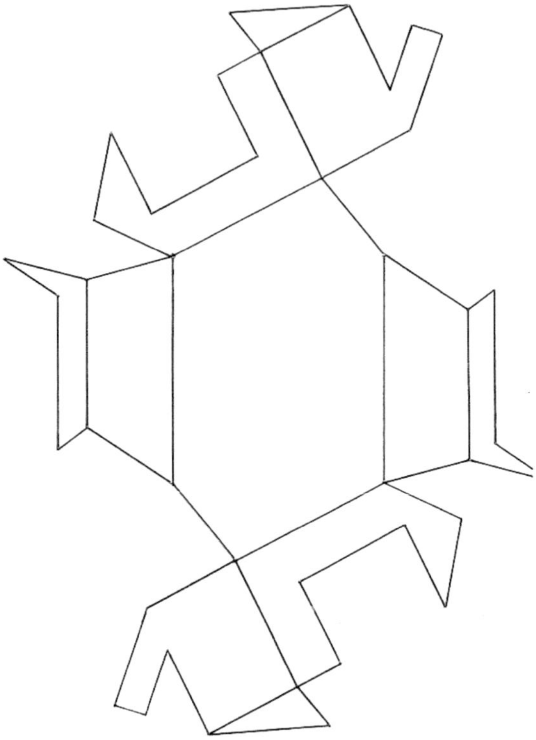

Fig. 44: The picture shows the paper sections of the feldspar twinned crystals (left).

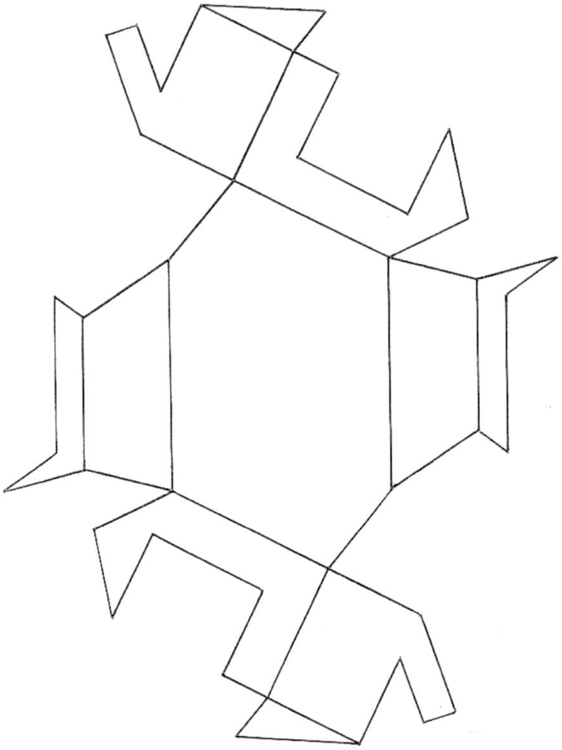

Fig. 45: The picture shows the paper sections of the feldspar twinned crystals (right).

10. The Shapes of Snow Crystals

Fig. 46: Wilson A. Bentley at his camera.

Fig. 47: Snow crystals from Bentley's book.

We can appreciate the infinite variety of snow crystals today because of Wilson A. Bentley's life's work (Fig. 46). He recalls January 15, 1885, when he was just 20 years old, as the greatest moment of his life. He had succeeded in taking the first successful micro-photograph of a snow crystal. This farmer's son from Jericho, Vermont, USA, was interested in all natural phenomena. His mother had already given him a microscope as a fifteen-year-old. He dedicated his further life to the photography of snow crystals in winter and the study of raindrops in summer. He became a pioneer in these two fields of science. It was not until 1898 that the first essay on snow crystals was published, and in November 1931, his work *Snow Crystals* [97] was published with 2,453 images, a selection from over 5,000 micro-photographs of snow and ice crystals. He died one month later! Many successors continued to work on the subject, but his work remains the unrivaled classic of images of snow crystals.

The images in Fig. 47 are arranged to show a sequence, from the completely confined crystals on the upper left to the lower right's radiating shapes. One can marvel at the symmetry and variety of forms that are all subject to an overall formative principle. It is astonishing that of the many billions, each has an individual shape different from every other! In contrast, each crystal has six branches containing an identical form, as do the right and left halves of each branch. It is exciting to follow how each crystal has been formed differently from the inside out, and yet all are subject to the same formative forces.

Figure 48 depicts three snow crystals from Bentley's book, from the middle of the third row in Fig. 47. Figure 49 shows the two balanced planes of symmetry that can be read in each snow crystal. Thin dashed lines show the three planes mirrored on each branch's right and left sides. Thickly dashed lines show the three planes mirroring the branches. Thus, six equal branches developed with twelve equal branch-halves. The angles are always the same: 60° between the branches and 120° between the branches' outer corners. There are four mirrors, but without a real side and a virtual side like a conventional mirror. Nevertheless, conventional mirrors are valid as an expression of the supersensible forces of form, representing each crystal's spiritual side. Due to the earthly side, which works from Earth's forces, an unlimited variety of individual shapes is created. Form and shape affect each mineral crystal. And all this happens in crystals that are only about 6 mm in size. The size does not play a role in the earthly world, which can be seen, for example, in insects.

Fig. 48: Three examples of snow crystals from Bentley's book, from the third row of Fig. 47.

Fig. 49: The two balanced planes of symmetry that can be read in each snow crystal.

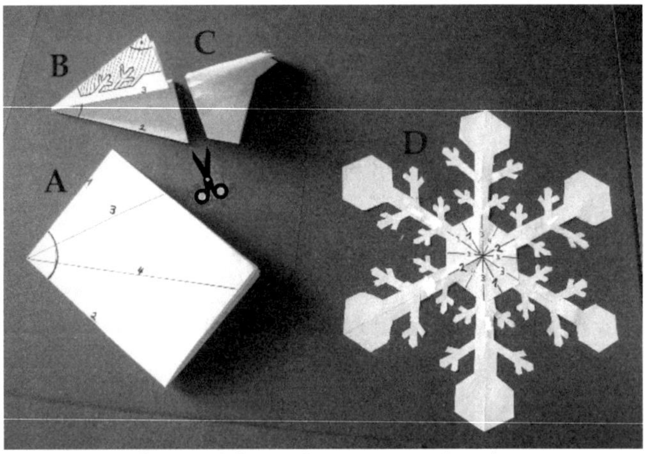

Fig. 50: The form and shape of a snow crystal can be cut out to visualize the crystals' mirrored planes better.

Figure 50 illustrates how through folding and cutting, one can experience a snow crystal's form and shape and better imagine the mirrored planes. You can experience this yourself when you take a larger sheet of paper (say 12 x 18 inches) and then fold and cut it to obtain a six-pointed paper star. To accomplish this, fold the sheet in the middle (Fig. 50A), then again in the middle of this fold (Fig. 50A), then fold this again at the middle into three isosceles triangles of 30° each, for example, one triangle to the back, the other to the front (Fig. 50B) and then cut away the tips where a right-angled triangle is formed (Fig. 50C). The side opposite to this is the hypotenuse. It will form the axis of three rays. Parallel to it, by two folds (Fig. 50A), draw the desired shape (the white area defined by the hatched area, Fig. 50B). Ensure the lines in Fig. 50B are parallel to the two folded edges, which you can see clearly as 120° when you unfold the piece. Cut out the hatched area in Fig. 50B. When unfolded, you obtain a six-rayed paper star, mirrored on four planes like a snow crystal (Fig. 50D).

All four elements must be involved in this interplay of formative forces. From the Earth, a very fine grain of dust is necessary as the nucleus of crystallization. I was amazed at how much dust is in the air when we once had to thaw freshly fallen snow far outside of Weimar in the post-war period because a pipeline had frozen. Sahara dust can color the snow a brownish color in southern Europe, even as far north as Germany. In an industrial city in 19th century England, it was once measured and estimated that more than a million pounds of coal dust were precipitated by snow in one winter. Water is another prerequisite because the humidity must exceed 100%, which means it must be supersaturated. Since warm air can absorb more water than cold air, the moisture condenses to fine water droplets and forms clouds as soon as it rises to higher, colder air layers. The heat diminishes, and the temperature drops to below 21°F there, and the moisture condenses on the dust particles and then crystallizes into snow crystals. In the air, the forms can float freely between point and circumference and take on individual shapes. Laboratory tests have shown which shapes form depending on temperature, air pressure, and humidity. At temperatures between 21°F and 14°F, they form within a few seconds, especially needles and hoar frost. At temperatures down to about minus 4°F with high humidity, they form radiating, flat shapes that crystallize more slowly. At about minus 31°F and low humidity – I.e., in the upper troposphere – fine sheets are formed, clearly demonstrating that the two-dimensional shapes form from the outside. When the Sun's rays refract on the outer surfaces of the snow crystals in high cirrus clouds, a colored halo with a diameter of 22.5° from the viewer is formed around the Sun, with a color sequence reversed from a rainbow: red on the inside towards the Sun and blue on the outside. Such halos appear on average 50 to 70 times every year. Some people have seen them only very rarely, or not at all because you have to look towards the Sun when high cirrus clouds are in the sky. When the snow crystals have reached a certain size, they fall downwards, thaw, and become raindrops. Snowflakes are then formed near the Earth's surface from individual snow crystals that have swirled together. Under 21°F, they can freeze again and become sleet kernels, but not snow crystals. Thus the forms that snow crystals reveal in their rhythm between crystallization and thawing are an expression of the temperature stratification, humidity, and air pressure in the Earth's atmosphere. Consequently, the snow crystals demonstrate exceptionally well what forms all crystals: they are infinitely varied middle links between earthly and cosmic formative forces. Goethe writes about this, saying: "The same atmospheric water unites in the open air during a severe frost to form six-branched crystals, which form in plates in such an elegant way that they guide the tenderly devoted observer to the almighty power and presence of God." [98]

The following quotations are a kind of dialogue between Goethe and Schiller, but they are at different places in their works and were only compiled by me as a conclusion:

Who divides the seemingly unchanging array
Invigorating its movements rhythmically?
Who calls the singular to universal blessings,
Where it beats in beautiful chords?

Johann Wolfgang Goethe

Wer teilt die fließend immergleiche Reihe
Belebend ab, dass sie sich rhythmisch regt?
Wer ruft das Einzelne zur allgemeinen Weihe,
Wo es in herrlichen Akkorden schlägt?

Johann Wolfgang Goethe

Only the body is suitable for those powers,
Weaving the darkness of destiny,
But far away from the forces of time,
The playmate of blissful natures,
Transforms above in the corridors of light,
The shape that is divine among gods.

Friedrich Schiller

Nur der Körper eignet jenen Mächten,
Die das dunkle Schicksal flechten,
Aber fern von jeder Zeitgewalt,
Die Gespielin seliger Naturen,
Wandelt oben in des Lichtes Fluren,
Göttlich unter Göttern die Gestalt.

Friedrich Schiller

11. The Majestic Spirit Might of Iron

A contribution to its evolution in Earth's history

O Man,
You form it for your service,
You reveal its material value
In many of your works.
Yet it will only bring you healing
When it reveals to you
Its majestic spirit might.

Rudolf Steiner [99]

What is this "majestic spirit might" of iron? How can we reach the inner essence of such a substance? When I want to learn about a person's true being, I ask for their biography. We can also approach the being of a mineral in the same way when we immerse ourselves in the biography of its substance. A mineral today is dead and crystallized. Nevertheless, it reveals itself in the stages of evolution inscribed in the Earth in the sequences of its deposits. A rock's spatial layering corresponds to its biography and the temporal succession to events in the Earth's evolution. Using the Goethean method, we can start from the perceptible phenomena to learn about the mineral formation process throughout the Earth's history. This way, an uninterrupted path of knowledge can lead to Rudolf Steiner's spiritual scientific findings.

The iron minerals in the Earth

Iron is by far the most common of the ore-forming metals. Iron minerals are found in almost all rocks. They penetrate all major formations of the Earth's crust, with quantities decreasing sharply from below upwards. In the Earth's interior, iron brings about the Earth's magnetism. At a depth of more than 30 kilometers, the upper mantle consisting mainly of green olivine (peridotite, chrysolite) is brought to the surface by volcanic eruptions from the basalt layer above it (Fig. 51). Basalt is hidden under the oceans and continents, and half of it consists of iron-rich minerals containing about 10% iron. Ore veins with iron minerals emerge from the granite layer in the continental crust above. In the gneiss and mica-schist, huge banded iron ore deposits are lenticularly embedded, consisting of thousands of fine layers of red iron ore and magnetite. The sand, limestone, and clay that follows contain only a small amount of mighty iron minerals, and these were mainly concentrated and secreted by the metabolic processes of bacteria. An example of this can be seen with the pyrites in Paleozoic limestone lenses, the red sandstone, and brown iron ores of the Mesozoic, especially the pyrite, which is finely distributed in many calcareous and slate deposits. Such phenomena suggest that the iron minerals were not formed by inorganic-chemical precipitation but can be explained as resulting from a colloidal-bacterial life process, as can still be observed today on algae deposits in the water, and with numerous rock formations during the evolution of the Earth organism.

Thus iron minerals in the Earth's history developed under quite different conditions than exist now, from life processes that are dead today, when the former life sphere has decayed into the air, water, and rocks. When the minerals reach the Earth's surface in today's air-water conditions, all the primary-formed rocks and minerals weather; the iron ores weather into water-bearing brown iron ore. This substance is further precipitated by bacteria and is converted into very stable iron oxides and color almost all of the rocks brownish or a reddish-yellow, including the brown soil. Thus, the iron ores in this overall sequence have undergone the most distinct qualitative and quantitative evolutionary process of all minerals throughout Earth's history, from which the details of their biography can be read (See examples in Fig. 52 through Fig. 61.

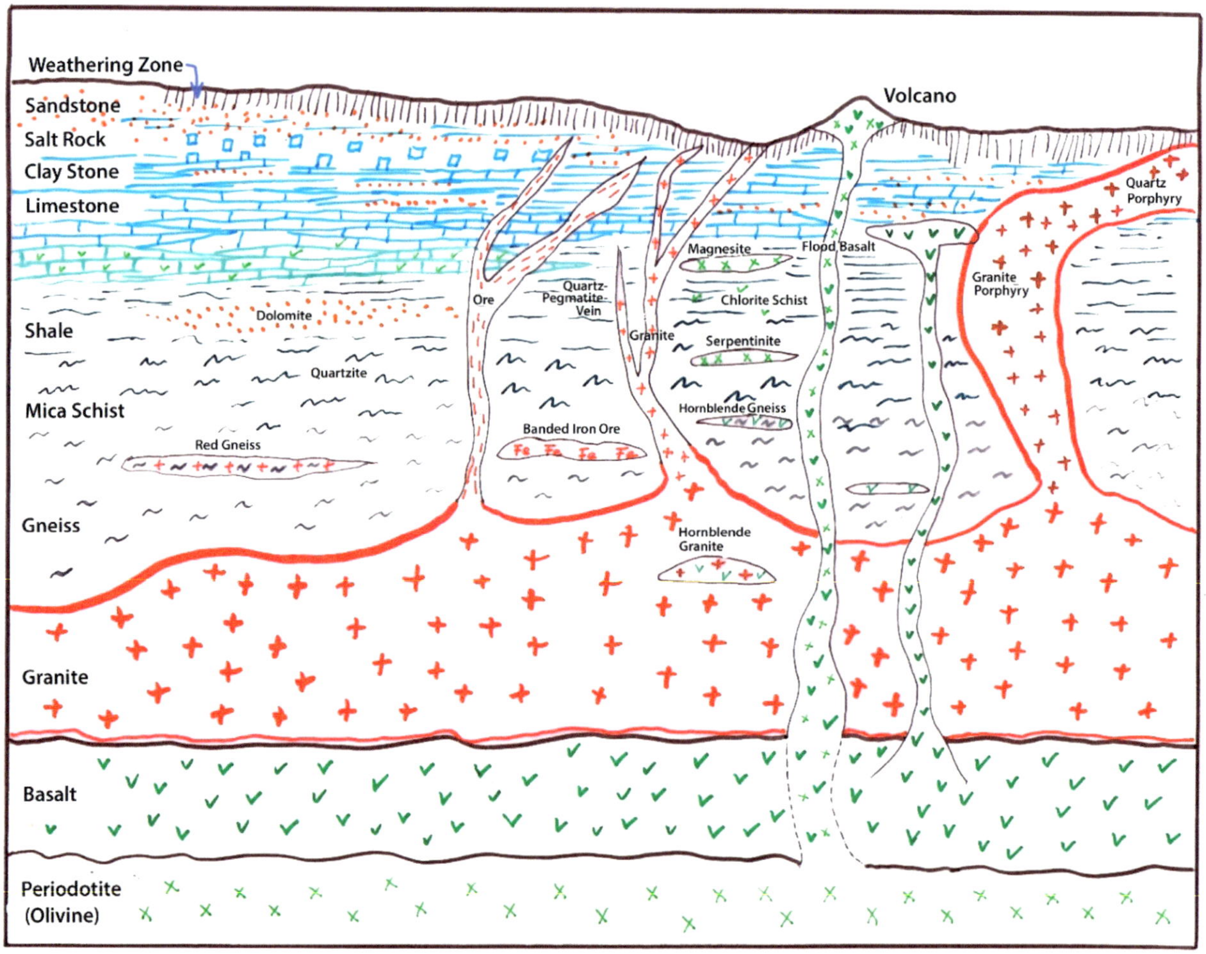

Fig. 51: This section through the Earth's crust shows the rock structure, a lens of banded iron ore, ore veins, and at the top, the weathered layer of the rocks.

Fig. 52: Brown iron weathering crust on limestone.

Fig. 53: Brown iron weathering crust on sandstone.

Fig. 54: Yellow iron ore in veins.

Fig. 55: Yellow iron ore in slate.

Fig. 56: Brown iron ore

Fig. 57: Brown iron ore

Fig. 58: Red iron ore

Fig. 59: Red iron ore

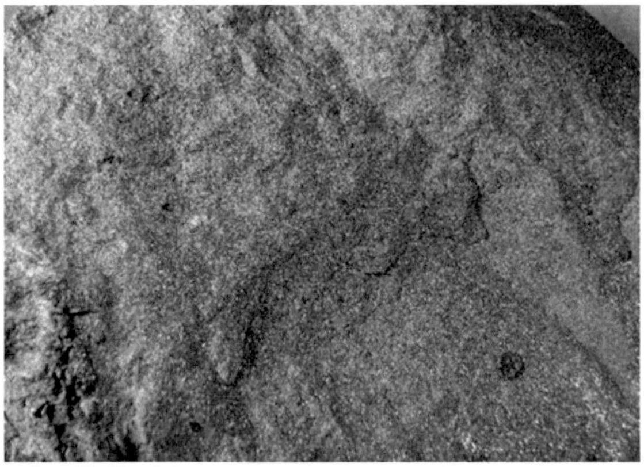
Fig. 60: Black iron ore

Fig. 61: Black iron ore

From iron ores to iron substance

Where was the iron during the Earth's early development, before it was bound and concentrated out of the fluidic air, hardened, and finally crystallized? The "origin" of iron ores can be traced by geological research back to when the mineral became crystalline. What happened before depends on the assumptions: Did the Earth originate from an exploded, radioactively heated dust vortex, or the life process indicated in many ways by the geological phenomena? Before the various substances could accumulate on Earth, processes must have taken place through which they "originated" in the true sense of the word – just as the living kingdoms of nature come from life. Rudolf Steiner put it like this: "[...] we must bear in mind that a substance is a process brought to a standstill, a frozen process, as it were. Properly speaking, we should not say pyrite, but pyrite-process." [100]

The oldest known substances of the Earth's mantle – the iron shell, the olivine, and basalt layer (blue-green, in Fig. 62) with its high iron content – was formed during the Archean period. The Earth has this in common with the solid planets and cosmic bodies. The further continental development from the granite layer on (red in Fig. 62) – covering about 30% of the Earth's surface – took place in the whole known cosmos only on the Earth, from granite to the origin of life and the human being, the water, atmosphere, and numerous other phenomena. These are excellently described by the Earth sciences, although their causes cannot be explained there. Rudolf Steiner's spiritual-scientific research continues beyond visible nature's limit of knowledge, through which a complete picture of evolution becomes possible.

Rudolf Steiner describes the separation of the planets within the solar system as a physical expression of the development of spiritual beings and human beings. The visible planets only mark the boundaries of their actual spheres of activity. [101] Thus the metallic has been brought about by the forces of the planets, the Sun, and the Moon, forming the seven principle metals in the Earth's orbit in a very finely distributed state [102]. The other metals related to the primary metals developed from the interactions of several planets. Such effects were probably only possible after the respective planets with their orbits had separated from their original relationship to the solar system, as happened during the Hyperborean and early Lemurian periods. From then on, the metals condensed from the imponderable into the airy-fluidic state of a life sphere. Rudolf Steiner called them "air-born" and "earthbound." The forces of sulfur and oxygen played a crucial role in this process. The sulfuric processes brought more the spiritual and the oxygen processes more the living-etheric into the Earth. In its ores and minerals, iron becomes bound with oxygen and sulfur, whereas other metals usually occur only in one or the other form of bonding.

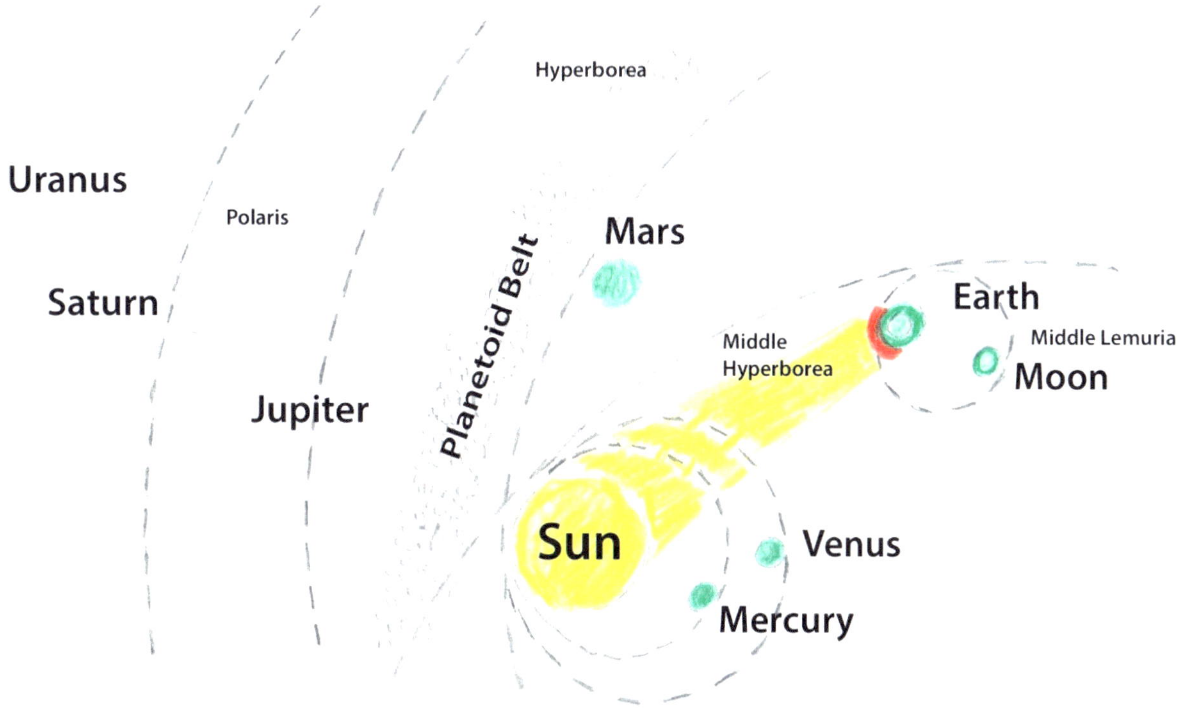

Fig. 62: The separation of the planetary bodies through the contraction of their spheres of activity. Rudolf Steiner sketched only the circles on the drawing during a lecture and wrote the names that designate the planets [103].

Between Jupiter and Mars orbits lies the planetoid belt (also known as the asteroid belt). The images from research probes show irregularly shaped rocks there with smoothed surfaces and impact craters. They could be attributed to fragments of broken planets. About 10,000 are known to be more than 0.6 miles in diameter; the total number of asteroids is estimated at one million. Rudolf Steiner describes the cause of this in such a way that retarded spirits of the second hierarchy fought with advanced beings from the same hierarchy: "Between the orbits of Mars and Jupiter, you have a host of small planetoids. These are the wreckage of the battlefield of the battle in Heaven that was fought between two points of Cosmic time when our Solar System first contracted as far as Jupiter, then to Mars." [104] Orbital disturbances among the planetoids make mass calculations possible, which can then be converted to density values. They show that about three-quarters of the planetoids correspond to the basalts, others to the olivine in the basalt, and only about 5% correspond to the density of the iron-nickel alloy, which is the best known.

Further, it can be concluded that the planetoid belt is structured according to the magnitude of its mineral masses. The objects in the belt's outer area correspond more to meteoric basaltic rock; in the middle belt area, those rich in olivine, and those inwards towards Mars resemble iron-nickel meteorites. From this, it can be assumed that the solid planets, the Moon, and the Earth, are constructed like the planetoid belt is organized. Thus, while the solar system contracted and left the planetary spheres behind, the Earth formed in the same way, so that among the oceans and continents, the outer cosmic past is mirrored. [105]

However, before the Earth's actual development began with the Polaris period, it went through earlier states, which Rudolf Steiner described in Theosophical terms (Fig. 63). At that time, the iron forces of Mars were already acting on the astral Earth: "Spatially, one has to imagine the Earth in its fourth round before it became ethereal, that is physical, and passed through Mars – which was ethereal at that time. Schematically one has to imagine it like this: This passage lasted even further into the physical Earth time; and while it was taking place, the leading beings took from Mars the desire-body (cama-materia) necessary for the intellectual soul, and since this has its physical vehicle in the warm blood, [...] the iron of Earth was inserted at that time as a component of the blood." [106]

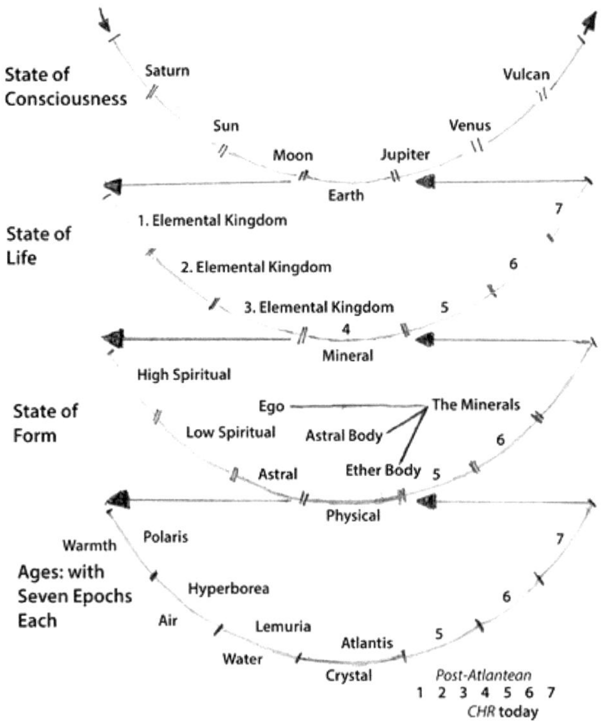

Fig. 63: The states of life and form, which from the Polaris period onward have brought about the Earth's mineral-physical development. [107]

In the physical development of the Earth that followed, the hyperborean age's hot-air Earth sphere was imprinted with the life body, thereby also the human being, repeating the “Old Sun” processes. The Sun sphere with the Earth-Moon body contracted into the Mars sphere. In the subsequent Lemurian epoch, the Earth-Moon condensed to a fluidic state in which the Moon began to separate. During this epoch, iron from the sphere of Mars was impressed into the Earth's body (Fig. 64). This Rudolf Steiner describes as follows: "At the time when the Moon was still separating from the Earth, a very mysterious process took place in our cosmic evolution, which is very difficult to explain and which is called the passage of Mars through our Earth. [...] Because, while the Earth was still connected with the Sun, the Mars mass was in it.” [108]

IV. Earth Round

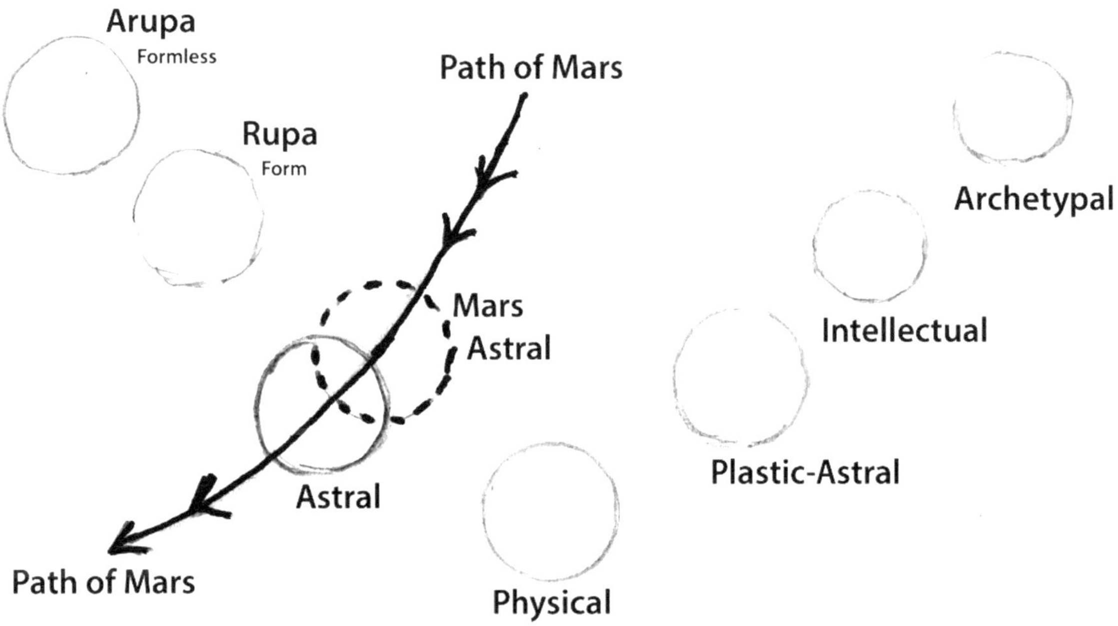

Fig. 64: Mars in the astral state passes through the etheric Earth during the fourth state of form.

The Sun then separated from the Earth, and after that, Mars separated from the Earth and left behind the substance known as iron." [109] Thus, above the iron-nickel layer of the Earth's lower mantle, the olivine layer and the basalt layer were formed. The solid planets Mars, Venus, Mercury, and the Moon (blue-green in Fig. 62) also formed. Together with the other cosmic bodies, the Earth underwent this common evolution and formed the basalt layer hidden beneath the continents and the oceans.

It was only on the Earth that the granite and gneiss crust separated from the hot-airy-fluidic life sphere – as well as the slates, limestone, and the salt rocks later on. They formed the continental crust on the side facing the Sun because the Earth had not yet started to rotate at that time (yellow in Fig. 62). It occupied only one-third of the Earth's surface and broke up into the individual continents at the end of the Paleozoic period. On the side facing away from the Sun, the basaltic masses were pulled together towards the Earth on the one hand and the other towards the lunar body, described as the "Moon separation." For the external process, the same principle could be assumed for the separation of the planetary bodies.

The common origin of iron and warm blood

Iron "was first present in an astral form, and then it condensed. If we trace the Earth back to that time, we find fewer and fewer warm-blooded animals. It was not until the middle of Lemurian times that warm blood appeared with the Mars impulse. The iron entered the blood at that time." [110] And elsewhere, it says: "At the time when the Earth and the Sun separated, the world body (the Earth) consisted of easy-flowing substances and was likewise permeated by fluid currents from the planet Mars. There was no trace of iron on the Earth before this passage of Mars. The iron resulted from this passage. All substances containing iron, like our blood, are under the influence of Mars. Mars has colored the substance of the Earth, and its influence has made the appearance of red blood possible." [111]

Rudolf Steiner shows that the metals in general, together with the human organs, were formed by planetary forces within the Earth's early life sphere. [112] Thus, their substances can have a healing effect on human beings today; they are, except for iron, hostile to life in a concentrated form. During the Earth's further evolution, all metallic substances were concentrated into a crystalline state; iron mainly in the basaltic, and the other planets' metals more into the granitic. Thus, in the early Lemurian period, the Earth's life sphere was largely free of metallic substances; it became detoxified. However, imponderable metallic effects remained in the Earth's orbit. And where today the metallic comes into contact with water and is dissolved, the corresponding bacteria immediately ensure that the substances precipitate again and are transferred into the ores' solid state (Figs. 52 and 53).

From a geological point of view, it is evident from the hardening of the crystalline rocks from the Archean to the end of the Paleozoic (Middle Lemuria) that mainly Moon forces were active. The gneiss layer is already partially eroded at the beginning of the Paleozoic. At the same time, the first hard-shelled fossils appeared. The granite layer underneath the gneiss hardened and crystallized later in the continental areas, until the end of the Paleozoic; only in the mountain building (orogeny) zones did it remain fluidic and soft until the Cenozoic. Also, the existing fauna became extinct at the end of the Paleozoic. The cold-blooded amphibians, the poikilotherm (temperature varying) reptiles with their upright tendency, and the very first small, warm-blooded mammals newly appear. In its ongoing appearance, the entire animal world reflects human evolution when you look at it as the earlier hardened members of the soft-bodied human being, which was still hovering in the upper part of the life sphere. These events also express what Rudolf Steiner describes as a Luciferic influence on the astral body and through which the ego was drawn deeper into the etheric and physical body. The warm, red blood formed the physical carrier of the ego-organization. "This blood was not in a human body before an ego incarnated, so this red human blood is related to the evolution of the Earth itself." [113] "Within the human organism, blood cells [...] form a system of forces that is oriented towards the external, physical nature. This, however, is continuously overcome by the ego-organization." [114] Or seen from another aspect: "Iron is that metal that constantly affects the organism in such a way that the astral body and ego do not connect too strongly with the physical and etheric bodies. We actually have a continuous therapy in the human being: iron therapy; and the moment a person carries too little of the iron, they immediately have the desire to become calm, to become flaccid; and as soon as the person develops too much iron, we have an involuntary activity, a fidgeting. Iron is the regulator of the connection between the physical body and etheric body on the one hand and astral body and ego-organization on the other hand." [115] Could it be that the iron in the blood reveals itself through the ego-forces "with its majestic spirit might" which can bring about a balance between the desire to become limp or restless? Wouldn't this be a picture for Michael with the scales and his sword of meteor swarms in the Michaelic Age?

Thus the sensory phenomena described by natural science can lead us directly to anthroposophy. "Natural science has led man to know one side of world evolution, for example, that out of lower animal organisms the higher, more perfect and so on up to the human being have emerged over time. [...] But that is only one side. The other side is that side that appears before our soul when we say to ourselves: The human being has had to develop from its original divine-human disposition [...] to their present form. Gradually he has had to cast out from himself, first the lower animals, then increasingly all the animal forms there. He has overcome this, set himself apart, cast it out. Through this, he has come to his original destination. [...] You see in that which shows you knowledge of nature, that which you must exclude from the knowledge of man. What does that mean? It means that the human being today must study natural science. Why? When a person looks into a microscope, then the person knows what spirit is not. When the person looks through the telescope into the far reaches of space, that is revealed to this person what is not spirit. [...] When one comes to realize the need for natural science to receive the thrust towards truth from this counter-image of truth, the strength will be developed that can symbolically indicate the conquering of Michael's dragon. [116]

12. Peat and Brown Coal – Graphite and Bituminous Coal – Petroleum in Sediments

Coal and its related formations are considered "biominerals," and in the case of bituminous coal, it is also referred to by mineral names: vitrinite = bright coal; durain = splint or dull coal; anthracite = carbon-rich bituminous coal; and graphite. They have fallen from a plant life process into the inorganic, just like the limestones or phosphates from animal life. If the Earth is considered an organism, all primary minerals in the rock formations would have to be considered biominerals, from the basalts to the clay slates and limestones. The differences would then only lie in the time of origin and the degree of hardening and crystallinity. In the narrower sense, all biominerals would be those exudations of the Earth's life sphere derived from plants, unicellular or bacterial life processes, specifically peat, brown coal, bituminous coal, anthracite, graphite, as well as crude oil, natural gas, and gas hydrate. In chemical terms, these would all be processes in which mainly carbon is bound from a gaseous state by oxygen and hydrogen into a fluid to solid state, accompanied by nitrogen, sulfur, phosphorus, and various trace elements such as iron, mercury, uranium, vanadium, arsenic, calcium, magnesium, and others, also depending on the weathered minerals in the environment at their formation, including quartz, clay and salt minerals, pyrite, and calcite, which in some cases have deposited in fissures in the bituminous coal, often as calcite or clay minerals.

Fig. 65: Peat cutting in Finland.

Brown and bituminous coals are derived from peat (Fig. 65). After dying, the plant cellulose remains underwater. The water cover prevents oxidation and bacterial decomposition processes from occurring: if a peat bog dries out, the plant substances it contains will quickly decompose. The higher plants cannot survive in deeper water. Thus, it must be assumed that shallow water depths over a wide area are sustained at the plants' growth rate, necessitated by the accretion of plant litter in the water column. The sandy or clayey layers that occur within the coal are explained as originating on the coast of a water body that repeatedly floods the coal-generating plant growth. But that is where the first problem with coal formation begins. With the thick seams extending over hundreds of kilometers, sea coasts no longer provide a realistic solution. Brown coals (Fig. 66) are usually up to a few tens of meters thick, bituminous coals a few meters. One meter of fresh peat is turned into 20 cm of mature peat by dewatering, and over time into 5 cm of brown coal. Suppose one continues in an uninterrupted "coalification process," which means developing an increasing carbon density over time. In that case, one will get 2 cm of bituminous coal and even less with anthracite and graphite. The average thickness of coal seams is between 50 m and 100 m. The thickest brown coal seam in the world in Morwell, Victoria, Australia is 320 m thick, which would correspond to 6,400 m of peat; the thickest bituminous coal seam is 200 m thick in Fushun, China, and would consequently require about 10,000 m of peat accumulation! These are peat thicknesses that cannot be explained geologically since plants only grow in very shallow water, even if one were to use large, continuously sinking basins and millions of years to "explain" this geologically.

Fig. 66: A brown (lignite) coal mine in the Lusatia region in eastern Germany. The huge machine on the right side with two excavator arms transports the excavated material across the pit from right to left. There can be seen the light-colored alluvial sands from the Ice Age that previously formed a layer above the pit on the right above. The yellow excavator digs up the coal in the middle, which is then transported on the long conveyor belt to briquette and heating plants.

The next problem is the conversion process from brown coal to bituminous coal. The peat is considered to have gradually turned into brown coal. The latter contains numerous plant remains, on the one hand, remains of wood (xylit) and large stumps; on the other, leaves and spores of trees that still exist today, especially in southeast Asia's swamp forests. Amber is also a fossil resin found in coal deposits (Fig. 67). The sands under the seams are bleached by organic acid, so brown coal reacts as an acid due to the humic acids it contains; bituminous coal, however, is alkaline! The transformation – the coalification – from peat to soft brown coal ends with a black, hard brown coal, also called pitch coal, which only chemically corresponds to brown coal, but externally and microscopically is already bituminous. This latter occurs only seldom and is formed at greater depth, where temperature and pressure are greater. Bituminous coals, however, are not further altered by pressure and higher temperature. Brown coal only loses its water through pressure, and higher temperatures due to contact with basalt do not produce bituminous coal but rather natural coke. Also, the geological time has hardly affected the coalification process for brown coals. There are a few older deposits of brown coal from the Lower Carboniferous, for instance, in the Moscow basin, and despite their age, they have not become bituminous coal.

Fig. 67: Brown coal with amber

Looking at the chemical analyses, it seems that brown coal is continuously changing into bituminous coal. The carbon content increases from 50% with wood in 10% steps to peat, brown coal, bituminous coal, anthracite, to graphite with almost 100% carbon. Water and carbon dioxide contents decrease accordingly. Nevertheless, brown coal and bituminous coal show essential differences concerning the degree of coalification and chemical composition. The decisive transition from pitch coal to the next stage of bituminous coal is rarely formed. This transition is referred to as a "coalification jump" in the continuity of the increase in carbon density. A bacterial biochemical process and dehydration take place in the transition from peat to brown coal. Bituminous coal is anhydrous and is subject to geochemical coalification, which releases methane and carbon dioxide. It is estimated that approximately 200 liters of methane can be produced from 1 kg of bituminous coal during further coalification. In mineral deposit science, brown coal is classified according to its water content, bituminous coal according to its gas content.

Brown coal and bituminous coal contain distinguishable plant remnants that are abundant in brown coal but not in bituminous coal! You can find the well-known, well-preserved leaf impressions and trunk pieces in the clay slates above the seams. The black bituminous coal usually consists of a fine-layered alternation of bright coal and dull coal (Fig. 68). The bright coal (vitrinite) makes up over 90% of the bituminous coal, which has no structure and rarely contains microscopically recognizable wood cells. Only in the dull coals (durain) can plant remains such as spores, pollen, leaf litter, or waxes be distinguished microscopically.

Fig. 68: Stratified bituminous coal.

Between bituminous coal and anthracite, which are both amorphous, and the geologically older graphite, which is crystalline, there is an additional phenomenological difference and another "coalification jump." It occurs in the oldest layers of the early Paleozoic period, in phyllite and alum shale. The oldest graphite appears in the Archean gneisses in South Africa. It also appears in Michigan and Greenland, starting in the Proterozoic period, and accompanies shale formations in the Paleozoic period.

Thus, a clear distinction can be made between three developmental series, which are limited by "coalification jumps": Peat to brown coal – bituminous coal to anthracite – and graphite. The geological formation periods overlap. The maxima are clearly separated: graphite in the Proterozoic and early Paleozoic, bituminous coal to anthracite at the end of the Paleozoic, and brown coal in the Tertiary, the Cenozoic period. All of the above-described differences and many more aim to question whether bituminous coal, anthracite, and graphite emerged from peat bogs through continuous coalification. From the aspect of the living Earth with its life sphere, different ideas may arise for their formation, connected with surrounding rock formations, which clarify the questions and continue directly to the limits of knowledge of coal geology.

A first plant-tinged, primary excretion process from the Earth's life sphere led in the gneisses to the crystallized graphite mineral. It originated from organic carbon, as isotope analyses show. A second excretion followed, which led from anthracite to bituminous coal. It ended in the non-crystalline, unstructured, bright coals and perhaps continued in places up to the pitch coal. The bituminous coals were formed by rhythmically assimilating processes from the Earth's life sphere, which can be compared to conditions found today in bacteria, protozoa, and lower plants. They formed soft-grained plant masses in which individual, structured cell tissues increasingly emerged, which had hardened most strongly, and in dull coals can be identified microscopically. From a geological point of view, these processes were connected with the primary substance separations of the mica series and ended with the first maximum of the bituminous coal formation in the clay schists at the end of the Paleozoic. Thus the plant world was individualized up to specific species. After only a few forerunners, it first appeared predominantly in the evergreen tree forms of the gymnosperms. They were surrounded by a hot fluid life sphere, as the hollow trunks of horsetail trees and their leaf formations show.

At about the same time, the Mesozoic saw the beginning of the third developmental series from peat to brown coal, which was completed at the beginning of the Cenozoic. Secondary rock formations mostly accompanied it. However, it led from individualized plants to coal and ran in the opposite direction to the first two developmental series! Thus it also corresponds to the secondary rock formation and biochemical coalification.

Yet another organic carbon process, the formation of crude oil and natural gas, ran parallel to secondary rock and coal formation and led to the original albuminous life sphere's demise. It also manifested in the exudation of huge quantities of oil, natural gas, and gas hydrates that were particularly evident at the transition from the Mesozoic to the Cenozoic. The problem arises again when considering how these substances could have been produced in such quantities and almost only during this time if one presupposes that the Earth is dead! The most recent formations are the gas hydrates that consist almost entirely of methane; they arose from life processes that can still be traced. Ideas can be developed from this about how the death of the Earth's proteinaceous life sphere could have happened.

Methane, which is gaseous under normal conditions, changes to the solid state at temperatures between 0° C and 15° C and pressures above about 50 bar. It forms a white, crystalline gas hydrate that immediately changes to the gaseous state under normal surface conditions. One-hundred and sixty-four cubic meters of methane become one cubic meter of gas hydrate. Mixed in are about 1% water, carbon dioxide, and ethane. These formation conditions are mainly submarine at the foot of many of the continental slopes between 400- and 1,500-meters depth and are also found in permafrost soils in Siberia and Canada. Today, methane in the oceans is mainly formed from the organic matter of single-celled plant organisms, the siliceous and calcareous algae. They float freely as clouds of intensive plant life in the oceans' surface layers, assimilating carbon dioxide from the atmosphere. The amount formed is estimated at 50 billion tons of carbon annually, or as much carbon as all the land plants on Earth put together. Such productivity is possible because phytoplankton has a life cycle of only a few days. After dying off, a portion of it then forms the organic part of the sediments. It is assumed that "bacterial decomposition of organic remains occurs in the absence of air" on the seafloor. Presumably, most gas hydrates have been formed geologically since the ice age. Loose sediments can absorb up to 10% gas hydrate. It is estimated that the total amount of gas hydrates stored in the loose sediments, when converted into tons of carbon, is more than twice as much as all other fossil fuels on Earth, meaning crude oil and natural gas and all carbon fuels together. However, the extraction of the gas hydrates is still technically challenging and involves risks.

Oil has been around since the Cambrian period. More than 80% of it was formed between the Jurassic and Tertiary periods, meaning at the end of the Mesozoic era and into the Cenozoic. This period was also when the huge oil shale deposits of Canada and Venezuela were formed. Like gas hydrates, the formation of crude oil and natural gas is attributed to protein- and fat-rich substances from plant and animal residues that were bacterially degraded in the absence of air and thereby reduced. In enclosed depressions containing stagnant water, digested sludge was formed, from which most of the so-called oil parent rocks (Fig. 69) were formed. However, numerous geochemical processes are still required before crude oil or natural gas can be produced from them. There are several complicated theories of origin because the origin of particular organic compounds remains unclear. Suppose plankton was the main source of oil and natural gas formations. In that case, larger deposits must have been formed in other geological periods, as is the case with gas hydrates. Seen from the premise of an airy, aqueous, protein-like life sphere, it becomes understandable because it finally died out at the end of the Mesozoic era and into the Cenozoic era. The actual cause of the origin of such unimaginable amounts of organic matter is mainly due to this dying off. The gas hydrates formed only the last large "remnant," while during the Ice Age, the oceans' water precipitated out of the atmosphere.

Fig. 69: Petroleum limestone from the Jura in Lower Saxony, Germany.

Rudolf Steiner describes these early substance processes in the following words: "[...] what geology presents can be understood only in connection with the living organic and spiritually permeated Earth. We do not have something primary in the geological formations we see, but we have something that is separated off, excreted. Actually, we have in bituminous coal formation only the simplest, most elementary process of mineralization. But everything we have in schistose formations, all crystalline formations, all this is excreted, is precipitated, is more or less what mineralized out of an original undifferentiated organic spiritual state." [117] And in another place, he said: "These imaginations had the form of giant plant-like organisms; and out of that which, as imaginations, formed within the earthly, there developed the later plant-like through the intake of the atmospheric substance, first in a volatile-fluid form in the circumference of the Earth. Only afterward did they sink down to the ground and develop into what is known to us as the plants. [...] And besides the silicic-acid-like, the calcareous was imbedded in this albuminous atmosphere in a fine distribution. And out of this calcareous again arose, from the coagulation of this albumen, the animal-like. And the human beings felt themselves within all of this. They felt that [...] they were one with the whole Earth. They lived in what formed itself in the Earth as plants through imaginations; they lived in what formed itself in the Earth as animal. Every human being felt essentially as spread out over the entire Earth, as one with the Earth." [118] And in other lectures, [119] he compares the formation of plants and carbon with measurable respiration. The human being absorbs oxygen from the air and gives off carbon dioxide. The plant absorbs the carbon dioxide and builds its body from the carbon, and gives off oxygen. Thus, you can understand the formation of peat and brown coal, crude oil and natural gas, and bituminous coal and graphite from the Earth's protein life sphere in a new way.

13. On the Essential Nature of Orthoclase and its Use in the Preparation of Horn Silica

Fig. 70: Orthoclase (reddish) and quartz (dark) have crystallized separately in this pegmatitic granite, the so-called "graphic" granite.

On the last day of the Agricultural Course, in the "Paths to Lost Forces Working in Nature" from June 17, 1924, Rudolf Steiner said: “Nature builds up a wisdom-bearing crystal. The wisdom-bearing crystal can delight us when we wander out into nature. But at the same time, we must be aware that gods had to die, not the death of the Earth, but the death of transformation; that is, the transition into the unconscious to come alive again in the shining crystal forms. And today, we must include it in our feeling so that when we look outward to what is dead, the life of the gods, which has been unconsciously resting in nature for thousands of years, shines through it to us. We have to find the possibility in our soul to feel and find this light which can touch us from the Sun and quicken our hearts as the divine everywhere in nature.” [120]

Questions and traditions regarding the horn silica preparation

During the discussion about fertilizer preparations, Rudolf Steiner said at the end of the 4th lecture of the Agriculture Course about the horn silica preparation: “Once more you take the horns of cows. However, this time you fill them not with manure but with quartz or silica or even orthoclase or felspar, ground to a fine mealy powder, of which you make a mush, say of the consistency of a very thin dough. You fill the horn with this.” [121]

The orthoclase is less well known and less frequently found than quartz, a particularly pure and perfectly formed rock crystal. For this reason, rock crystal is often used for the preparation. Would Rudolf Steiner have said quartz or orthoclase instead of rock crystal if he had meant rock crystal? There are two additional legends to the horn silica preparation:

Ehrenfried Pfeiffer shares in the Appendix to the *Agriculture Course*: “For the silica preparation “501,” Dr. Steiner said it would even suffice to mingle and knead a piece of quartz the size of a bean with soil from the land which is afterward to be sprinkled and put this mixture into the horn. This would already contain sufficient silica-radiation if a little of it was dissolved and stirred.” [122]

There is a letter (presumably in the archives at the Goetheanum) in which (presumably) Günther Wachsmuth writes to Count Kayserlingk on July 10, 1924, about a follow-up discussion with Rudolf Steiner: “At the time, the doctor stated that cow horns should be filled with finely ground orthoclase-feldspar, silica. When we were discussing where we could get a larger quantity of it, Dr. Steiner said that it was perfectly sufficient to mix a medium-sized crystal of this type, in a finely ground state, with some of the soil that would later be treated with this substance, in other words with finely ground crystal flour.”

One could think of "crystal" as rock crystal; however, quartz and orthoclase are also crystalline. No authentic source is known about the use of rock crystal. This situation gives rise to questions that are to be addressed in this paper: What is the relationship between orthoclase and quartz in the geological formation process? Which substantive forces within the common evolution of the living Earth and human being are we referring to in the sense of anthroposophy, not in the sense of a "rock cycle"? What other orthoclase-quartz rocks are still existing and what does "quartz or silica” mean in this context? What are the differences between quartz and rock crystal? Of what significance is the red color of orthoclase? How are orthoclase, quartz, and rock crystal characterized concerning the horn silica preparation? What does the outer and material similarity between orthoclase and potassium salt say? Since the feldspars change into clay minerals during weathering, the orthoclase questions are followed by those concerning clay preparation.

Mineral and chemical elements

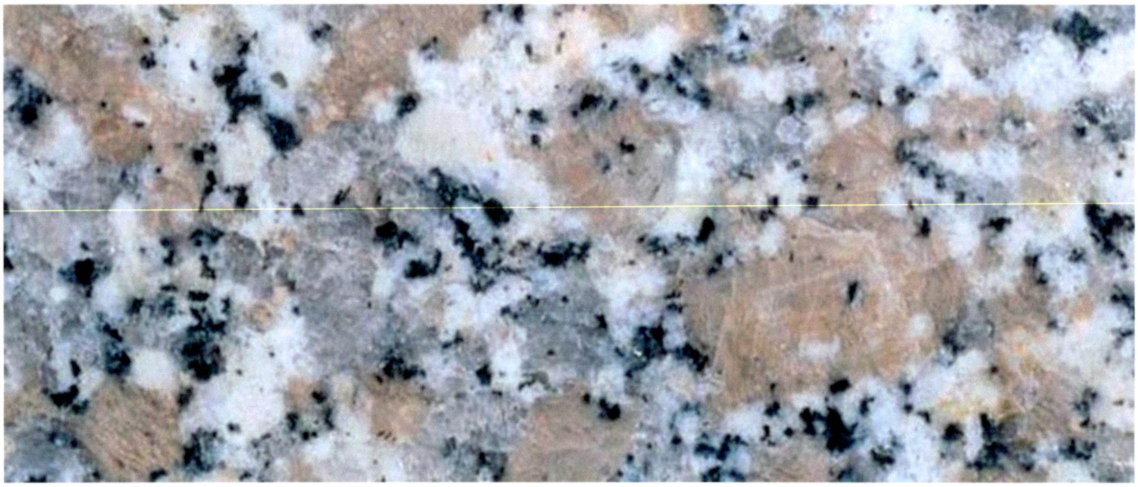

Fig. 71: A granite with a roughly balanced ratio of mica (black), plagioclase (white), orthoclase (red), and quartz (grey).

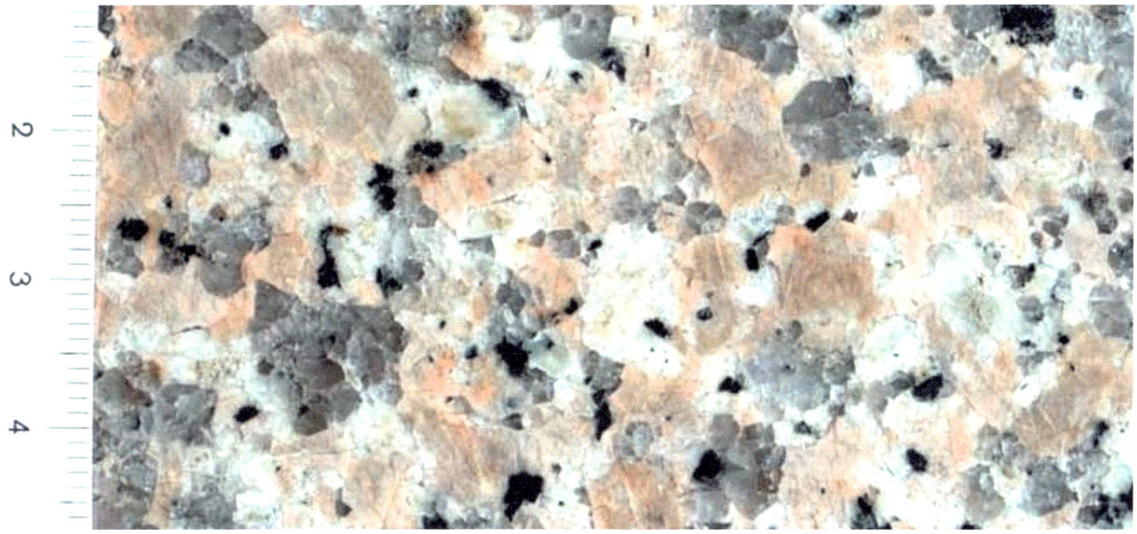

Fig. 72: A granite with a roughly balanced ratio of mica (black), plagioclase (white), orthoclase (red), and quartz (grey).

Just as we can get closer to a person's ego through his or her biography, it is also possible to understand something of a mineral substance's essence by its history of formation through its connection with Earth's evolution. But how can you recognize the history of formation from the phenomena of a solidified rock? Suppose you look at a granite (Figs. 71 and 72). In that case, you can often recognize the temporal succession of their separation and crystallization from the spatial juxtaposition of the grains: the mica often has a sharply defined form, which shows that they separated first from a previous fluid or soft state; the two feldspars – plagioclase and orthoclase – have already more or less merged into each other, and the quartz fills the remaining gaps and is the last to crystallize. Thus, from the juxtaposition, one can imagine the separation from an original totality. While the entire mass hardens, the earlier crystalizing mineral forms its shape better than the later one. This separation in time is a basic principle of every development in the living and mineral world. Although it is so, people continue to say that rocks are "assembled" from minerals. Likewise, a human being is not assembled from cells.

The actual process of becoming is incomprehensible as long as the formulations and concepts lead to misconceptions. This way of thinking and expressing originate from the 19th century when technology developed an artificial world. It was then that natural materials were first disassembled and combined from their parts to form new materials having particular properties. Minerals were also analyzed this way, whereby all their characteristics disappear, and new basic materials appear that are designated as chemical elements. They have completely different properties. They are completely artificial products, just like their new compounds, which do not occur in this way in nature. Thus, the orthoclase is called a "compound" of sodium, potassium, aluminum, silicon, oxygen, etc. However, it cannot be "combined" from these elements in the laboratory. In nature, the orthoclase has a completely different biography.

The chemical elements can characterize the minerals by pointing to something they have in common and comparable in their properties. For example, mica, feldspar, and quartz are called silicates because when they disappear in chemical analysis, silicon and oxygen appear in a ratio of 1:2 (SiO_2) next to other elements. In the laboratory, this is anhydrite of the artificial silicic acid (H_4SiO_4). However, in the natural context, we find neither the silicon nor the silicic acid or its anhydrite, but the crystalline quartz, the microcrystalline chalcedony, and the amorphous opal, which, of course, did not result from the dehydration of the artificial orthosilicic acid! I want to join Manfred von Mackensen [123], who begins with silica as an umbrella term for totality and describes the forming process as a silica process, which starts in the mineral, breaks down into the quartz and the numerous silicates, and can finally end in the technical laboratory product silicic acid or silicon (for instance, in the semiconductor industry). The words should be used in this sense. In the *Agricultural Course*, Rudolf Steiner usually speaks of siliceous and silica, silicic acid, and silicon. Looking at the connection between these statements, one has the impression that he uses these words mainly as synonyms to focus and refer to the silica process as a whole.

Feldspar and quartz in the rocks of the Earth's crust

Just as the chronological sequence of minerals can be read from their juxtaposition in granite, the biographical development of the Earth and its substances can be derived from the sequence of rock layers. Fig. 75 shows a general section through the Earth's crust, especially the rocks that pertain to the development of the orthoclase. It is one of the main minerals in the mighty crystalline basement rocks of the continental crust, mainly in granite and gneiss. We easily recognize it by its soft reddish to light orange color that results from microscopically fine particles of iron oxide (hematite) that have separated (only 0.2%). The orthoclase is an "alkali feldspar" (analytically appearing as potassium + sodium). It crystallizes as a single monoclinic crystal (ortho = straight, upright; clase = fracture) but usually forms mirror-image twin crystals (Fig. 73). The orthoclase can also be whitish, making it difficult to distinguish from the other feldspar, the white to light blue plagioclase. There are also numerous transitions between the two. The plagioclase is a "lime-soda feldspar" (analytically appearing as calcium + sodium). It crystallizes triclinic (plagio = oblique). The feldspars show clearly reflecting cleavage surfaces in a fresh fracture, after which they were named "spar." The angular difference between monoclinic and triclinic (straight and oblique) amounts to only 4° and is related to the Z-axis. In this respect, all four of feldspars' names refer to properties that only appear in chemical or microscopic analysis. It would be sufficient to speak of the orthoclase as red feldspar and plagioclase as white feldspar when considered from the directly perceptible phenomenon. The hardness of mica (2), feldspar (6), and quartz (7) corresponds to the sequence of separation; especially the quartz, which, as the last one to separate, goes most strongly into hardening.

Fig. 73: Orthoclase twin crystals in quartz porphyry and as models of the crystal form.

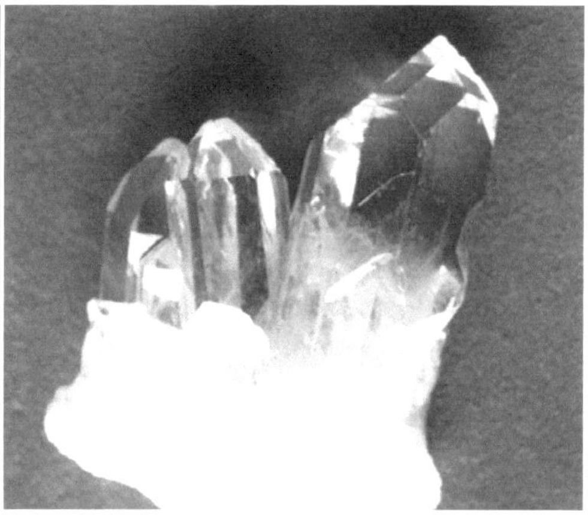

Fig. 74: The developmental step from quartz (bottom) to rock crystal (top).

As single or twin crystals, the orthoclase crystals have only retained their monoclinic shape on the outside. On the inside, during the slow hardening of the granitic rock, the original monoclinic structure crystallized into a less symmetrical triclinic structure. This orthoclase is then called microcline. The sodium feldspar separates from the potassium feldspar as microscopically fine, flat triclinic flakes and the quartz in fine filaments. The latter can also be observed on a larger scale with the so-called "graphic granites" (Fig. 70). Here, the feldspar and quartz substances are simultaneously separated and crystallized together so that their crystal forms are exactly parallel to each other. These "graphic granites" already show that quartz can be understood as a special feature of the silica-containing feldspar. This feature will be examined in more detail in the next section.

The rock sequence of the continental crust – covering one-third of the Earth's surface – will be roughly outlined here, as far as it is relevant for the development of orthoclase and quartz (Fig. 75). The mighty continental basement consists of granitic rocks and gneisses. Below them lies the basalt layer, which forms the Earth's crust as a hidden spherical shell under the continents and the oceans. In the basaltic rocks, the pyroxenes have developed as precursors of the mica minerals and separated from the white feldspars, the plagioclase minerals. Above this in continental areas follows the granite layer. In its deeper parts, the plagioclase continues to exist in the diorite and granodiorite; and the pyroxenes become the actual hornblende and mica minerals, which contain chemically bound water. The quartz increasingly appears towards the top. Thus, silica, as an independent mineral, gains a quantitative predominance. In terms of minerals, this characterizes the actual beginning of the Earth's evolution, which is in early Lemuria in the spiritual scientific sense. [124] Only in the granite layer's upper areas do the quartz and the reddish orthoclase feldspar dominate, next to plagioclase and mica. In the orthoclase, potassium appears first, tinged with trivalent red iron oxide. In the basaltic minerals pyroxene and black mica, the iron is still bivalent, i.e., it is bound to less oxygen. In all known planetary systems, the development of free quartz and water content has only occurred on Earth's continents. The Moon, Venus, Mercury, Mars, and other celestial bodies have only developed to the basalt layer stage. With the further development beyond pyroxene and plagioclase, the actual red orthoclase rich granite can be considered as the first and deepest rock of Earth development in a narrower sense.

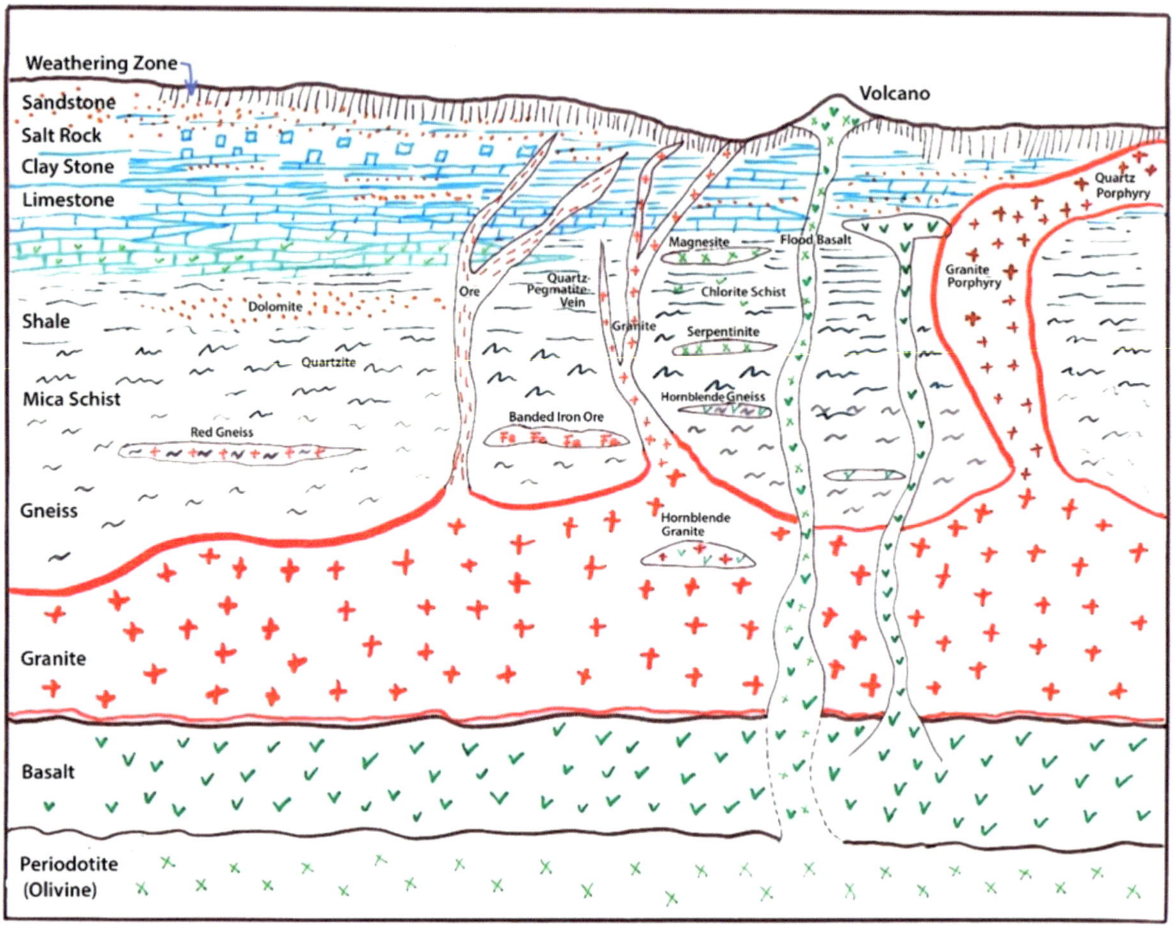

Fig. 75: General, schematized section through the Earth's crust, regarding orthoclase development.

The question about where the excess of quartz or potassium in granite comes from, or even water, air, and the richness of living nature, is difficult to answer if, as is usual in geology, a "rock cycle" is assumed where the geological processes of the now-dead Earth are used as a starting point to understand its history of formation. But suppose one starts from the life processes observed today. In that case, the formation of the rocks can be regarded as a separation from an original life sphere, as it can be derived in detail from the geological phenomena. The developmental processes are reflected in the rocks' sequence, which ultimately characterizes the evolution of nature's kingdoms and the human being, as Rudolf Steiner presents it. [125] From this point of view, an overall picture is possible today, to which the geosciences have compiled the phenomena comprehensively. [126]

Evolutionary paths of feldspar and quartz

The feldspars, with all their variations and derivations, are the most extensive group of rock-forming minerals. They can also be regarded as the original middle group. The dark, micaceous minerals, with their iron-magnesium content, separated from the basalt layer at the beginning of the Earth's evolution. From the granite, the evolution of the feldspars can be followed in several directions. Only those minerals and rocks, which are directly relevant as substances for preparation production, will be considered here. A granite specimen with large orthoclase-feldspar crystals (red), in between small quartz spheres (gray), and a few mica crystals (black) is shown in Fig. 76. The rock sequence's main development leads beyond the granites worldwide into the crystalline schists. The numerous gneiss varieties begin partly with the light, fine-grained granulite, a quartz orthoclase rock; they dominate the grey gneisses with much mica, white feldspar, and quartz, in which the lenticular red gneisses and the reddish feldspar orthoclase are embedded. The gneisses rise upwards into mica-slate, phyllite, and clay-slate. The feldspars increasingly disappear in this progression. What remains is a quartz-mica rock in which calcium silicates and dense crystalline limestones appear instead of feldspars. Roughly summarized, the younger limestones and salt rock formations follow the later substantive evolutionary stages of this series.

Fig. 76: A granite specimen with large orthoclase-feldspar crystals (red), in between small quartz spheres (gray), and a few mica crystals (black).

While numerous other minerals emerged from the feldspar evolution, quartz accompanied this development unchanged to the clay schists. In our context, these rock sequences can be seen in detail as secretions of life and evolutionary processes in the Earth's periphery, not as geochemical transformations within "metamorphic" rocks. With the clay slates, the quartz's primary evolution ends and is partially replaced by the sand brought in from the weathering of earlier crystalline rocks. The Earth's crust hardened and crystallized after the gneiss was deposited and opened the way for surficial weathering and erosion. Quartz, the hardest mineral in granite and gneiss, is left behind as sand, which is chalky or clayey and then becomes sandstone. The Paleozoic period often contains unweathered grains of feldspar called arkose. When the sand grains are combined with a siliceous binding agent, quartzite is formed, embedded in the crystalline slates in a lenticular form. The early quartzites, in particular, are colored red by fine hematite crystals – similar to the orthoclase. We have a pure siliceous rock in quartzite: fine quartz grains cemented by microcrystalline quartz, the chalcedony.

Another direction of feldspar evolution can be traced directly out of the granite, where the granitic mass further separated in the marginal areas of the granite massifs. The rock is very light in color since it consists mostly of quartz and feldspar (microcline). The dark mica had already separated from the feldspar in deeper areas. This granite is then very coarse-grained and is called pegmatite. Such pegmatites can be further traced in veins that have penetrated the overlying crystalline schists (Fig. 75). If they remain fine-grained, they are called aplite. The evolutionary continuation is followed by the ore veins, in which the quartz accompanies the entire process with only little change until it finally flows into quartz veins. Instead of the orthoclase, calcite spars appear in the upper part of the veins. So the feldspar finally merges here into the calcareous.

Yet another evolution also originates directly from the granite. The Earth's crust was so hardened at the end of the Paleozoic era that it fractured in places during the Variscan mountain forming in the Carboniferous period (Fig. 80). In the process, soft granitic masses penetrated the crystalline slates. Where it rose to the Earth’s surface, the granite solidified as thick slabs of quartz porphyry (=Paleo-rhyolite, Fig. 75, far right). This quartz porphyry is a reddish, fine-grained quartz-orthoclase rock, sometimes with larger orthoclase crystals (Figs. 73 and 77). It was formed in a final separation from the granitic, with about 50 % orthoclase, 40 % quartz, and only 10 % mica and plagioclase., Feldspar's evolution to its free crystal form (Fig. 73) can be observed in these granitic slabs. If gas chambers remained in the granite, quartz veins, or quartz porphyry, so-called druses were formed in which remaining gaseous substances further separated from the liquid-solid state. In them, the separation of feldspar and quartz continued.

The orthoclase remained substantially closer to its original state: at best, it became transparent and, with a hardness of 6, stands between mica (2) and quartz (7). The crystallization has further formed the orthoclase outside as twin crystals and filled the inside with cleavage planes. The right-left mirrored double shape of the twins is also closer to its origin when seen from the sphere's wholeness than the further removed crystallization of the quartz in the right or left forms. A conchoidal fracture inside the quartz indicates that hardening predominates here. The quartz's substance, which is usually opaque white due to fluidic inclusions, reaches its highest grade in rock crystal, where the columns can form freely all around (Fig. 74). Thus the rock crystal stands substantially at the very end of the natural development of silica. Only in the laboratory can the separation of silicon and oxygen be continued. The individual being of a crystallized mineral in its purest form manifests in the orthoclase twins and the rock crystals. This purity is perhaps what appeals to us about rock crystal and what the Greeks experienced when they called it the "Krystallos" in the true sense of the word. It is perhaps what causes us pain when we have to break it up for the preparation.

Fig. 77: Quartz porphyry with plagioclase (white), quartz (grey), and orthoclase (red) in fine-grained orthoclase quartz matrix.

A completely different development of the siliceous minerals began with the clay-slate. We are now reaching the secondary rock formations whose substances are no longer primarily deposited from the Earth's life sphere but instead originate from animals' life processes or the erosion of primary rocks. Thus, bulbous or stratified silica deposits are often found in the limestones, called siliceous shale, radiolarite, flint, or chert. The silica was enriched by life processes of plant-like unicellular organisms, radiolarians, or silica sponges from the surrounding water and embedded in colloidal form. After the animals' death, the silica dissolved, rose with the pore water in the calcareous mud, flocculated, and hardened, so the crystals in the colorless chalcedony remained microscopically small. Occasionally the nodules are also colored black by organic substances, as is the case with flint.

The potassium salts as a polarity to the orthoclase

In general, after the separation of the crystalline shales, the feldspar substance within the Earth's life sphere changed into the calcareous and then into salts (Fig. 75). The salt rocks are the last primary substance separated from the life sphere. They were concentrated in the fluidic, mainly through bacterial activity, and were deposited in countless rhythms, especially in the Permian, at the end of the Paleozoic. They accumulated in huge continental salt basins with a length of several hundred kilometers and a thickness estimated at approximately 1,200 meters. Salt deposits of such large dimensions cannot be observed today. This amount of salt cannot be derived from the weathering of feldspars: and it has different proportions than those from evaporated seawater or the salt lakes. Many questions here remain unanswered for academic science.

The sequence of the salt minerals is essential for our purpose here. In general, the rhythms begin above thin layers of mudstones. The limestones with dolomite [$CaMg(CO_3)_2$] and calcite [$CaCO_3$] come next. They change into anhydrite [$CaSO_4$] or gypsum [$CaSO_4 \cdot 2H_2O$]. Then follow the salts or evaporites: first rock salt [NaCl], and then the potassium salts in the broader sense (Fig. 79):

Carnallite [$KMgCl_3 \cdot 6H_2O$], kieserite [$Mg(SO_4 \cdot H_2O)$], kainite [$KMg(SO_4)Cl \cdot 3H_2O$], sylvite [KCL], polyhalite [$K_2Ca_2Mg(SO_4)_4 \cdot 2H_2O$], bischofite [$MgCl_2 \cdot 6H_2O$], in manifold mixtures and alternating layers ("hard salt"). Suppose one starts from the Earth's life sphere. In that case, it is possible to recognize a progression in these last precipitations of calcium-salt rocks, which in substance correspond to what separated from the granite early on. Both can be compared through the chemical elements that characterize the minerals. Thus, the salts' basic development is as follows: dolomite → limestone → gypsum → salt rock → potassium salt or, chemically speaking: calcium → sodium → potassium, plus magnesium. This series corresponds to the feldspar sequence from plagioclase (calcium → sodium) to orthoclase (sodium → potassium). Another characteristic is interesting: some potassium salts are slightly reddish due to small hematite content (less than 1%) and resemble potassium feldspar in color (Figs. 78 and 79). This phenomenological and chemical parallelism concerning the development of feldspars at the beginning of primary precipitation, with the development of salt at the end, is quite astonishing! However, while the feldspars solidified to a solid, crystalline structure, the salts did not go so far into hardening – hardness 2 to 3 – and tectonically, they even react plastically.

These aspects seem to be important for fertilization, resulting *in the Earth's vitalization,* [127] whereas mineral fertilizers with potassium salts *only support growth* by activating wateriness, *not through the Earth's vitalization* [128]. Rudolf Steiner's concept of fertilization is completely different from that of conventional agriculture. Mineral fertilization should *increase the supply of nutrients to meet the cultivated plants' nutrient demands effectively.* [129] Justus Liebig (1803-1873) started his experiments on artificial fertilization in 1840. In 1843, carnallite was found in two shafts in Stassfurt, with which he carried out his experiments; from 1861, potassium salt mining began and, starting in the 20th century, had a tremendous boom worldwide. In 2002, twenty-seven million tons of potash were mined worldwide, mainly sylvite and kainite, containing potassium in water-soluble form and kieserite because of the magnesium. However, Rudolf Steiner uses yarrow in a homeopathic way to counter the potassium and sulfur forces. [130] He responds with orthoclase as a substance that originates from when the silica and potassium forces first appeared together in connection with the Earth's development; the potassium fertilization is based on the polar opposite substance, salts that were the last mineral precipitation from the dying life sphere.

Fig. 78: Granite with orthoclase (red), quartz (grey), and some mica (black).

Fig. 79: A red and white potassium salt.

Questions and traditions regarding the clay preparation

Further questions that are related to the orthoclase pertain to the clay preparation. In the 2nd lecture of the Agricultural Course in 1924, it says: "Everything in the nature of clay is, in reality, a means of transporting the influences of cosmic entities in the soil from below upward." And shortly afterward: "[...] if we are dealing with a soil which does not readily or of its own accord carry upward the influences which should be working upward in this winter season, then it is well to add a dose of clay to the soil. (I shall indicate the proper dose later on)." [131] However, Rudolf Steiner did not specify this dosage, and nobody asked for it. This omission leads us to the question of clay preparation. "In the middle of winter, in January and February, let the water freeze in the ground. Add clay. It will drive away the pests." [132] Further notes [133] are related to the etheric, whereby the clay stands between sand and calcium: "The clay carries what is above in the air as light and warmth into the soil." This would be the countermovement to the idea "from below upwards." However, it is difficult and ambiguous to draw conclusions from these notes because Rudolf Steiner did not give details in his lectures.

There are also traditions about clay preparation. Ehrenfried Pfeiffer, in the Supplement to the Agricultural Course proceedings, refers only to quartz with the horn silica preparation. He says nothing about a specific clay preparation. Now we learn from Nikolaus Remer: "An extension of the horn-silica preparation: In the Agricultural Course, Rudolf Steiner stated that it would be good to add a little clay to sandy soils in November in a dosage that he wanted to specify. But this did not happen. In 1955 I asked Dr. E. Pfeiffer about it, and he informed me that Rudolf Steiner had subsequently given him the relevant information. Orthoclase has to be finely ground. A bean-sized quantity is to be mixed with the soil of the field to be treated, poured into a cow horn, buried in the field over the summer, and then sprayed in November like horn silica." [134] Martin W. Pfeiffer wrote in 1975 [135] that the information about a clay preparation was unfortunately not given! Nikolaus Remer only informed me in 1983 that Ehrenfried Pfeiffer's reference in the 1929 edition of the Agricultural Course was incomplete.

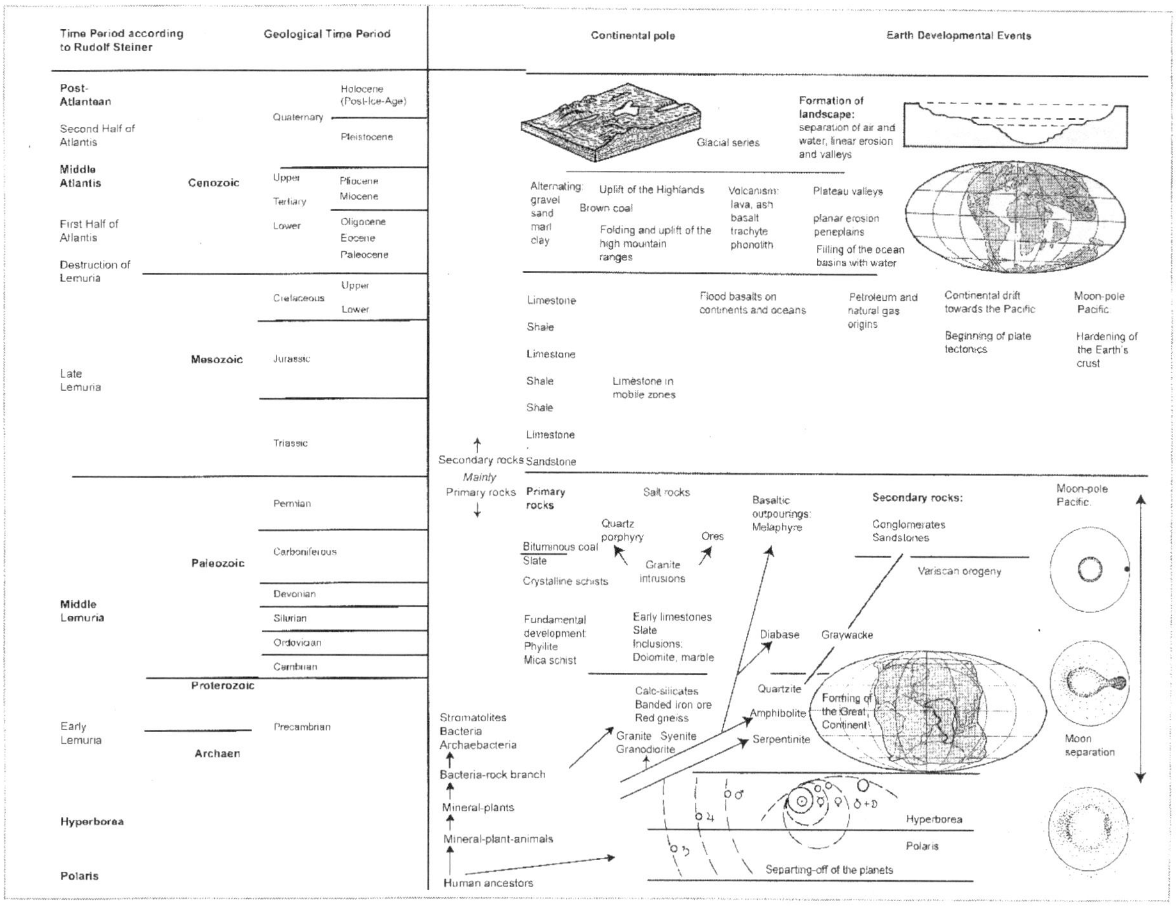

Fig. 80: Period, Geological Scale and Earth Developmental Events [136]

Here too, questions arise: Is this clay preparation the one Rudolf Steiner meant when he announced it in the second lecture? Have three factors been connected here after a lengthy period of 25 years: the statement from the 4th lecture that one could also use orthoclase; plus the additional message that one should mix it with field soil; plus the fact that certain clay minerals are formed when orthoclase has weathered? Rudolf Steiner did not mention orthoclase together with the clay but with quartz. Is it possible that clay is meant as a substance for clay preparation, and only the information on dosage and application is missing? In my opinion, the authenticity of this statement by Nikolaus Remer about the clay preparation remains open. In light of these suggestions, however, various attempts have since been made with clay preparations (Franz Lippert, Greg Willis, Hugh Lovel, Raimund Remer, Walter Stappung), of which Walter Stappung [137] provides a good overview. However, the results are difficult to grasp.

The formation of clay minerals

The minerals of crystalline rocks originated from Earth's life sphere processes under very different conditions than those prevailing on the Earth's surface today. In a way, the primary rock formation falls into timeless duration due to crystallization.

The water that filled the ocean basins is the last thing that separated from the life sphere. Warm air remained behind. Thus, a new development began for the existing minerals when the rocks are exposed to weather, fresh water, and air, called weathering. All crystalline minerals today are actually – speaking physically and chemically – "unstable systems." Rock decay and erosion have existed since the Earth's primeval times, since the middle Proterozoic (early to middle Lemuria). The primary Lemurian formation zones became increasingly smaller. All surface rocks were then subject to the completely different Atlantean and Post-Atlantean conditions. [138]

During weathering and depending on the environmental situation, most minerals are dissolved or transformed by air and slightly acidic water. The primary separation process that formed quartz, feldspar, and mica in the granite led to new secondary minerals. Of course, the formation phases can also be active multiple times: forming, reshaping, and newly forming in alternating rhythms – more or less quickly, depending on their inner stability. The quartz is only slightly soluble in an alkaline environment. The separated silica precipitates again as quartzite in an acidic environment, as happened largely in the Tertiary period (Fig. 80). With the mica and feldspars, we get into the realm of chemical transformations, which mostly lead to the clay minerals. These are very fine-grained, crystallized, flaky, layered-like-mica, flexible, rolled, and plastic. The silica part remains the defining basis. Water and salts can be chemically bound or deposited between the crystal layers and then released again; the clay minerals "expand." The water can carry away other soluble substances from the source minerals. However, no ions or atoms combine to form new minerals; instead, the primary minerals continuously turn into the secondary ones. One does not taste the sodium, potassium, or chloride ions, although you taste the salt content and see the iron content in the soil's brown color.

Feldspar, for example, the orthoclase (microcline): $K(AlSi_3O_8)$ has 70% silica, 16% clay, and 14% alkali (mostly potassium), as well as traces of red iron and calcium. In tropical climates, it becomes kaolinite: $Al_4[(OH)_8Si_4O_{10}]$ is formed when two-thirds of the silica and the alkali content is dissolved and carried away, 14% chemically bound water is added; the clay Earth portion (Al_2O_3) remains, as does one-third of the quartz as sand. This newly formed unconsolidated rock is known as kaolinite. It is the most stable of all the clay minerals. It requires abundant water for its formation, specifically a humid, tropical climate, which prevailed worldwide during the Tertiary period (Fig. 80). Most kaolinites were formed at that time when the air and water sphere began to separate. Independent of this, humus soils were also created as a new, plant-animal-mineral whole during this period., It is possible to imagine these soils as a transformed, last remnant of this life sphere when seen as a process. Under current conditions, similar life processes occur as during the primary rock formation.

Another developmental path for clay minerals originates mainly with dark mica biotite, hornblende, and similar minerals. They lose iron, magnesium, and alkalis when they weather. The light-colored mica muscovite weathers much less readily, loses potassium and can absorb water. It hydrolyses, swells, and is partially transformed into hydromuscovite; it is transformed further into illite in a moderate climate. In an alkaline environment, montmorillonite is formed and can contain 12–24% water. While kaolinite is also formed in humid climates and cannot store as much water, montmorillonite – sometimes called smectite – is formed mainly from rocks containing high calcium-magnesium-iron contents that are not easily washed away. Here the alkalis can chemically bind water. Montmorillonite is the main clay mineral in bentonite clay and is capable of considerable swelling and then rereleasing the water. Transitions are frequently formed with mixed-layer clays: muscovite-hydromuscovite-illite-montmorillonite, which shows how a continuous process occurs. Transitions make it difficult to identify and name the different stages.

All clay minerals remain in a colloidal state, are variable, and not as thoroughly crystallized as the orthoclase. The silica, iron, and salt components also remain colloidal in them, which is the state the orthoclase substance was in before it crystallized. All fluids in plants, animals, and humans are in a colloidal state into which the etheric body can penetrate. As a result, the humus and clay minerals are variable, responsive, and changeable substances, and consequently closer to the life processes. This development can also perish. It is kaolinite in particular that from the orthoclase forms the most stable clay material. A further mineralization stage is formed by bauxite: a solid, microcrystalline clay rock with abundant gibbsite $Al(OH)_3$ and brown iron (iron oxide, Fe^{+3}), which, depending on the subsoil, is bound to silica or calcium. Kaolinite and bauxite were mainly formed in the Tertiary and have been preserved since then.

The silica and clay minerals as preparation substances

To sum up, several parent rock materials could be used in the horn-silica preparation. Orthoclase originated in the younger reddish granites, which characterize Earth's evolution in a specific sense with the anthroposophical perspective of a living Earth. To that belongs the abundance of quartz, the step from plagioclase to orthoclase, thus from calcium to sodium and potassium. Thus, the orthoclase is a silica substance that anticipates a salt-(potassium)-process and is a counterimage to the potassium salt. It characterizes the step from calcium to the salts as the precipitation of what has been carried on in the Earth's life sphere by animal being; to the first rudiments of thinking in humans during the middle of the Lemuria; to the potassium-to-sodium process in humans. [139] The orthoclase stands between the silica and the calcium and shows by its clay content, which appears through weathering, its relationship to clay. Compared to quartz, it remains substantially closer to the original state of the unity of silica, clay, and salt forces, like the ego. Likewise, the formative forces are retained in the microcline so that the monoclinic has not completely fallen into the triclinic. The laminar crystallization appears on the outside and inside, in the case of quartz, only on the outside. The subtle amount of hematite also points to the step from bivalent to trivalent iron at the beginning of the Earth's development, which in mineral terms substantially corresponds to hemoglobin. The orthoclase is one of the most common minerals worldwide and is widespread, especially in upper areas of the granite massifs, separately in pegmatite and aplite veins or graphic granite, as well as in red gneiss and also quartz porphyry.

The quartz is first contained in the Earth's crust as the siliceous base of the silicate minerals. With the separation of the granite from the life sphere, the silica substance appears in abundance. In this sense, it "originated" in early Lemuria, at the time when the human sense organs were created. This fact is the reason for quartz's healing effect on the sense organs of humans and plants. Within the granite, the siliceous quartz separated from the feldspar as a mineral in its own right. Thus the quartz is even more widespread than orthoclase. As the last one to be separated from the granitic, it went more strongly into hardening, shown by the missing inner structure and greater hardness. In the orthoclase-quartz-pegmatites, it was further separated from the quartz veins. As chalcedony, it remains micro-crystalline, as in agate or jasper, which could develop within quartz porphyries. In the druses, the quartz can reach its highest individualization in the quartz rock crystal. Quartz is not only found in the granite and gneiss layers but also in the sedimentary rocks mentioned above:

- Sandstone (quartz grains with a calcareous or clayey binder)
- Quartzite (quartz grains with a siliceous binder)
- Arkose (quartz grains with feldspar grains)
- Chalcedony from organic processes in flint, chert, radiolarite, and siliceous shale

The essential differences in the range of variation between orthoclase, quartz, and rock crystal must be considered for the horn silica preparation. In terms of substance, it seems that this is a matter of processes and not about ego-character or beauty. With its original and more universal, balanced composition, the orthoclase seems to be more suitable for this preparation than quartz, which has fallen more strongly into hardening. In rock crystal, the silica reaches its most specialized end state, to which all crystalline minerals can more or less attain. To my knowledge, Rudolf Steiner did not make a statement about the use of rock crystal as a preparation substance. Its use instead of *quartz, silica, or orthoclase* can be understood in clear chemical terms but does not consider the essence of its substantive biography in Earth's history.

When processing the orthoclase, one should try to choose the purest red feldspar from the granite and, when crushing it, one should try to pick out the mica flakes; a certain amount of quartz would not be a problem. Conversely, if working with quartz, an orthoclase component would not be a hindrance. Rudolf Steiner indicates (Agricultural Course) that a mortar should be used. With quartz, because of its hardness, this would mean using an iron mortar and then an agate mortar, which may not be available. During the subsequent grinding on a glass plate (hardness 5), it must be understood that a considerable amount of potassium/sodium and some other materials from the glass would also be mixed in. With the orthoclase, these amounts would have been obtained naturally, directly from the mineral! During further mixing with water, the silica is dissolved more easily by the alkalis in the direction of a colloidal state, the so-called "water glass." The orthoclase is easier to crush because of its hardness of six and its strong cleavage surfaces.

Concerning a biodynamic clay preparation, a question would be to what extent something can be discerned in the described processes of clay formation, in terms of its forces, and could it be *"a means of promoting the cosmic forces in the soil from bottom to top*, for such a soil *that does not easily carry up by itself what is supposed to work upwards in the wintertime* [...]." [140] This concept is not only about the physical and chemical characteristics where the clay minerals absorb water and alkalis and then rerelease them. Instead, shouldn't one perhaps think about the forces of separation that transform the solid feldspar and mica minerals? As shown above, they disappear with all their properties through weathering, and, instead, the clay minerals appear with completely new properties. They remain substantially in the siliceous; the calcareous-salty feldspar and mica minerals have been separated. The new totality of the humus is created in the aqueous colloids by the vitality of the bacteria.

What does it mean to *add some clay* to the soil? [141] This idea can be understood as a material amount that is worked into a sandy soil to retain water better; a smaller amount added through compost, an addition to the preparation, or even finer processing as the preparation itself. To understand the orthoclase as a substance for a clay preparation might seem obvious because a part of the feldspar is transformed into clay minerals when it is weathered, especially when it is used to develop a field, and above all, when it transforms into the kaolinite, which has already been characterized as more stable. If one also thinks about how suddenly Rudolf Steiner conceived of the compost preparations (who would think of dandelions instead of orthoclase when talking about silica and potassium?), [142] the connection between orthoclase and clay seems to me to be inadequately grasped for clay preparation. Also, the tradition is ambivalent.

In conclusion, Rudolf Steiner characterizes in very apt words the soul-spiritual qualities of silica, lime, and clay: "The silica-nature is the universal sense within the earthly realm, the limestone-nature is the universal craving, and the clay mediates between the two. Clay stands rather nearer to the siliceous nature, but it still mediates towards the limestone. This should be pursued so that one comes to an experiential knowing." [143] This brief description of the development of the silica would like to lead to this kind of experiential knowledge of the forces of these three substances. A closer understanding of the mineral entities is possible through the geological biography of the substances. In this simplified presentation, I want to provide a basis of judgment for the reader; in itself, the processes are much more complicated. I do not want to advise on where and how the substances are to be used. Usage is a question of practical experience regarding whether one or the other substance appears more suitable under certain soil, climatic, or landscape conditions. Fully characterizing practical experience and proper usage is something that will be left to further experimentation and research.

14. Time-Concepts in the Earth's History

Relative Times in Rhythms

Events in human history are ordered in years. We also need a similar measure for the Earth's history, but simply giving an age in years does not say much. As a geologist, one is frequently asked about the age of a rock. On a geological time scale, every number remains unimaginable and abstract with its millions of years, a pure calculation quantity. Today, geological processes can no longer be directly observed or modeled in the laboratory in a way that remains true to nature, such as with the formation of mountains, new animal species, or the granite in the Earth's crust. The time factor appears as the limit of knowledge. What is it about rocks that indicates temporal processes to us at all? What reality do time measurements in millions of years have for our powers of comprehension? How does a five-million-year-old rock differ from one that is one-fifty or five-hundred million years old? What is there in a rock that points to processes in time? We encounter something in a rock that is unchanging and permanent; then again, we find images of diverse rhythms of color, form, and substance. We experience rhythms in all life-processes, heartbeat, breathing, and the Sun's daily and yearly course. Rhythms testify to developmental steps that go continually forward as they oscillate within a polarity. The successive layering of strata also points to relative time-sequences, ever younger from below upwards, as was pointed out by Dane N. Stenson in 1669. [144]

The year is a cosmic rhythm deriving from the interactions between Sun and Earth; it forms, for example, the yearly rings in trees. These have been counted, measured, and compared using the rhythms of thickness to give dendrochronology, reaching back almost without gaps to 5,000 BC. [145] With oaks from river gravels, one can go back even further. One can apply varve-chronology for the post-ice-age period, counting and comparing fine layers in clay deposited by melting glaciers. In recent years, interesting results have come from borings in continental ice in Greenland and the Antarctic, where over 400,000 rhythmic layers of ice can be correlated using isotopes of oxygen. These layers are understood as yearly rhythms that are interpreted as variations in ocean temperature. The rhythm of cold periods during the ice age is also preserved in the calcareous shells of unicellular organisms from deep-sea sediments. For earlier geological epochs, numerous other rhythms in sedimentary layers have been researched and tabulated, as with the fine sandy clays of the Tertiary, the lime-clay deposits of the Mesozoic, in greywacke of the Lower Carboniferous, and the millimeter-thick layers of anhydrite from the Zechstein, which can be paralleled over a certain distance. Annual rhythms are thought to be the cause of these observations. Of course, this is only an approximate 'absolute time' because the Earth's orbital period around the Sun and the Earth's rotational rate have also evolved to their present rhythms.

The further we go back in the Earth's history, the more we have to reckon with other causes of rhythmic life processes. From an anthroposophical perspective, the day and year as successive cosmic occurrences began in the second half of Hyperborea and the Lemurian epoch's beginning (Fig. 81). This fact is shown by comparing crystalline schist rock formations with the spiritual scientific research of Rudolf Steiner. It can be mentioned here that only during the Atlantean epoch, perhaps around the middle, did the year come to approximate its present length. This time would correspond geologically to the end of the Tertiary.

A large part of geological research is concerned with the time-relationships between rock sequences. Rock layers showing continuous deposition can be separated by boundaries where a clear period of erosion has occurred. During this latter, other Earth-historical events have happened, such as mountain building or animal evolution. Determining the age of rock would determine how long it has rested unchanged in the Earth's crust by comparing it with other developmental processes that bridge between its origin and the present. Temporal changes in a rock's evolution always remain external and are not internalized as with an organism, so it is possible to speak of age. Only in the context of the whole Earth is it an internal process. Each rock preserves the history of the Earth's crust in its location. Geology compares sequences of such phenomena and creates a comprehensive picture of the relationships between individual sequences. Through this work, the pictures indicate differing time-structures, rhythms, and speeds of development. Thus the succession can be read, whereby the explanations remain within the geological phenomena to which radiometrically dated events can be assigned.

The most important method for determining the relative age relationships between rocks is their connection with the evolution of animals and plants. Today stratification sequences can be profiled worldwide and placed in sequence, often down to a meter's precision, which is routinely done by the petroleum industry using microfossils. This work provides a detailed overall picture of rock formations worldwide. Nevertheless, there are difficulties – rocks with the same fauna were not necessarily deposited at the same time, rock types of a given epoch vary spatially, there are often large spatial gaps between exposures, different species of animals extend their range at varying rates, and each has its specific habitat.

The rocks of the Earth's crust can thus be impressed with a relative time-signature in three ways:

1. Through the succession of strata from below upwards,
2. Through the impression of rhythms in the rock itself, and,
3. Through the evolutionary steps of life as shown by the fossils they contain.

There are thus three different sequences of events, each related to the others. Besides, we have the various yearly rhythms for the recent past and radiometric dating using the various radioactive decay series for isotopes contained in particular minerals. Based on this, a concrete concept of time for the Earth's history becomes applicable, as Rudolf Steiner already developed in his youth.

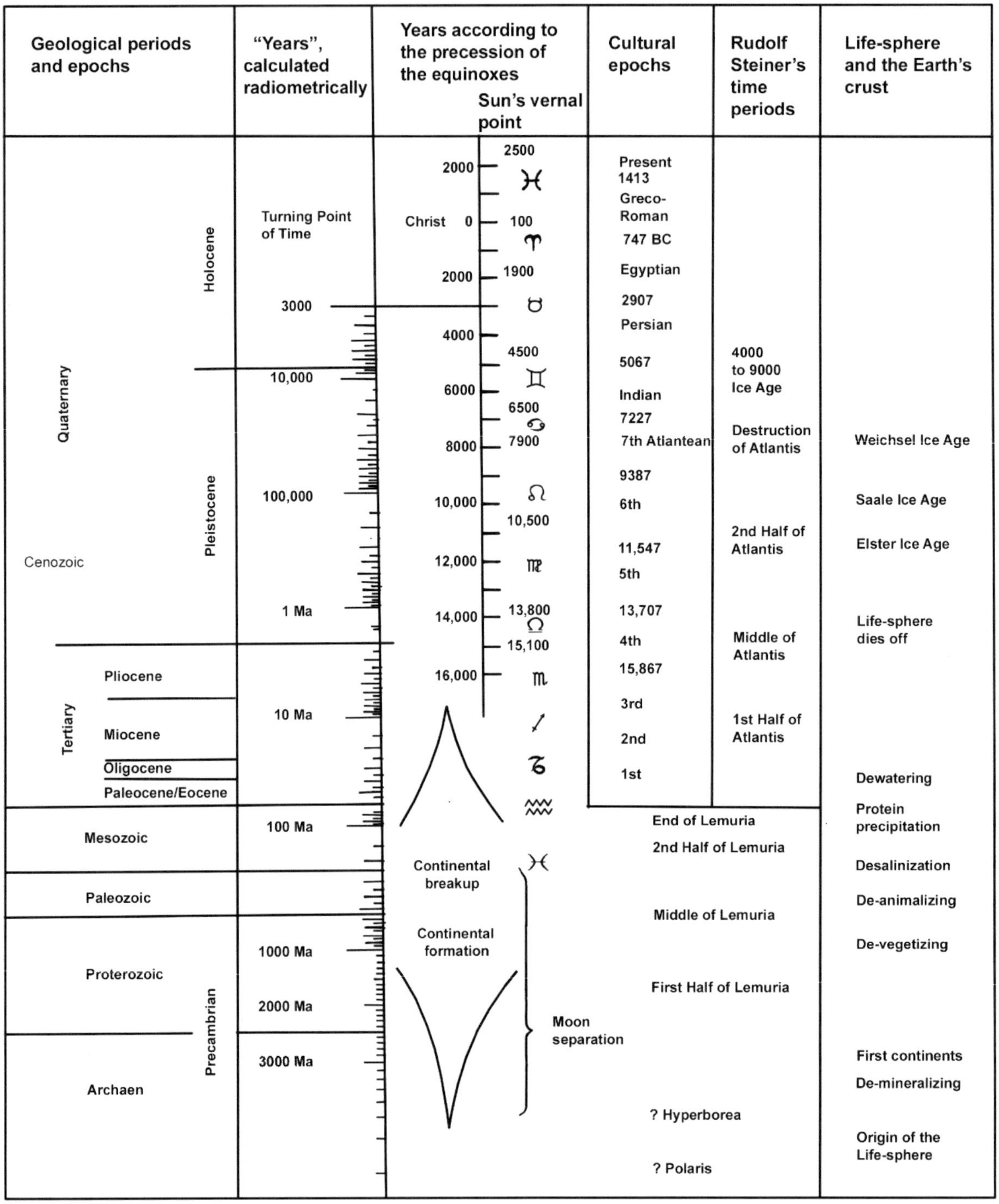

Fig. 81: The natural- and spiritual-scientific arrangement of the Earth's history of development, as presented in the book The Mutual Evolution of Earth and Humanity. A sketch of a Geology and Paleontology of the Living Earth. Rudolf Steiner described the development of Earth and human being. [146]

In present-day geology, time is thought of in the Newtonian sense. Accordingly, it runs independently and outside of things. It thus considers radioactive decay rates as providing an absolute measure. In contrast to this, Rudolf Steiner wrote [147] that there is neither an absolute space as an empty vacuum nor an absolute time as an unchangeable measure of successive events. Time is an expression because events are dependent on each other, according to their content, in a particular sequence. Time applies only to processes in the sense-perceptible world; it manifests when a being comes into the world of appearance. The being itself, on the contrary, can only be grasped as an idea. It remains spiritual, in the realm of duration. There is indeed evolution in the spiritual also, but this follows different laws. We can experience time because we swing continually from the soul-experiences bound up with the time-dependent world to the world of duration experienced during the day in thinking and at night in sleep. As early as 1882, Rudolf Steiner wrote of the importance for the future of science that the Newtonian time-concept be corrected. For the Earth's history, this would mean that one would work towards a new, more real concept of time by seeing it as the expression of sequences of inwardly dependent events and facts. This method will be applied in the last section of this chapter, Patterns of Geological Events, after the possibilities and limits of absolute time-determination have been considered.

Absolute Time-Determination

Efforts to determine Earth-historical times using years began with the scientific age. In 1648 the Anglo-Irish prelate of the Anglican church, James Ussher, using Biblical references, calculated that the divine creation began on Sunday, October 23 in 4004 BC, around 9 o'clock. Towards the end of the eighteenth century, the Frenchman Buffon calculated the Earth's age as roughly 75,000 years, based on a metal sphere's cooling rate. In 1862 Lord Kelvin used similar methods in England, came to 25 million years using a starting temperature of 1,000º C, or up to 200 million years using other temperatures. In 1902 Rutherford and Soddy found a new heat source in radioactivity, and with this calculated a cooling period of 400 million years. Today the accepted value has increased tenfold, using extrapolation based on radioactive decay rates.

The most important method for determining the absolute time for Earth-historical events is various decay series of radioactive isotopes. Time values in years can be calculated from the experimentally determined decay rates for isotope atoms, the 'half-life,' and by chemical analysis of a sample to determine the relative quantity of undecayed isotopes to decay products. Regarding a linear timescale, the starting-point quantity of original isotope atoms decreases exponentially (Fig. 81: "Years," calculated radiometrically; the decay curve of a radioactive isotope). Said differently – the quantity of not-yet-decayed substance increases exponentially as one goes back in time. The half-life values measured are assumed to be constant; the physical laws affecting them have not changed since the crystallization of the oldest Archean gneisses.

The decay rates are specific to the given isotopes; some are extremely slow, others much more rapid. For example, uranium 238, important for geological use, has a half-life of 4.4 billion years, decaying to thorium 234. The latter's half-life is only 24.1 days, changing to Protactinium 234 with a half-life of 6.7 hours, then to Uranium 234 with a half-life of 250,000 years, then to Thorium 230, and so on. The end-product is lead 206, which is not radioactive and is thus stable. Other decay series involve similarly slow rates (Fig. 81).

The geologic times calculated using these decay series give the impression that Earth-historical processes took place very slowly at earlier stages, measurable in millions or billions of years, and have become increasingly rapid. They yield a great age for the Earth and its first life-forms. However, the larger these values are, the less meaning they convey. They become increasingly a mere number, used in practice to arrange events relative to one another. In the sense of Rudolf Steiner's time-concept, the numbers form another series of facts which, according to their content, are related to other facts such as rock layering. In this sense, they have become a useful tool, allowing, for example, the worldwide correlation of fossil-free Precambrian rocks, or determining the hardening time of igneous bodies, placing them in connection with Earth-historical epochs.

Using radiometric dating, it has also been possible during recent decades to research the rhythmic reversal of the Earth's magnetic pole in more recent geological epochs. The magnetite of the ocean floor basalt preserves the orientation of the Earth's magnetic field at the time of crystallization. Along with the overlying sediment layers, this information allows for a correlation of the ocean floor's age with the timescale based on index fossils.

In all of these cases, geology compares sequences of facts related by their content: the crystallization of minerals in rocks, isotope decay, the magnetic field, fossils, or continental movements. However, geology does not use results indicating millions of years only to indicate relative age. They are also used, for example, to conclude the rates of sedimentation, continental movement, erosion, the time involved in the origin and extinction of animal species, etc. Far-reaching conclusions are drawn concerning the course of evolution.

At the beginning of every physical-etheric organic development, there occur relatively more developmental steps, metabolism, and growth than occurs later in the same amount of time. Later, all processes slow down with age until death – although with a rhythmic variation. The bivalve in Fig. 82, which is used as a representative for many organic developments, grows more rapidly and then more slowly. The growth-lines show how with each year, less substance is added. The growth along the apex, for the first eight years is (in cm) 6.1, 2.8, 1.6, 1.0, 0.6, 0.5, 0.4, 0.3. On the one hand, other animal and plant groups show the same principle in their early development and on the other in their specialization. Different building-plans appear at first in rapid evolutionary steps, which later become more differentiated only in details. A human being's developmental steps are measured embryonically in days and weeks, for an infant in months and a child in years.

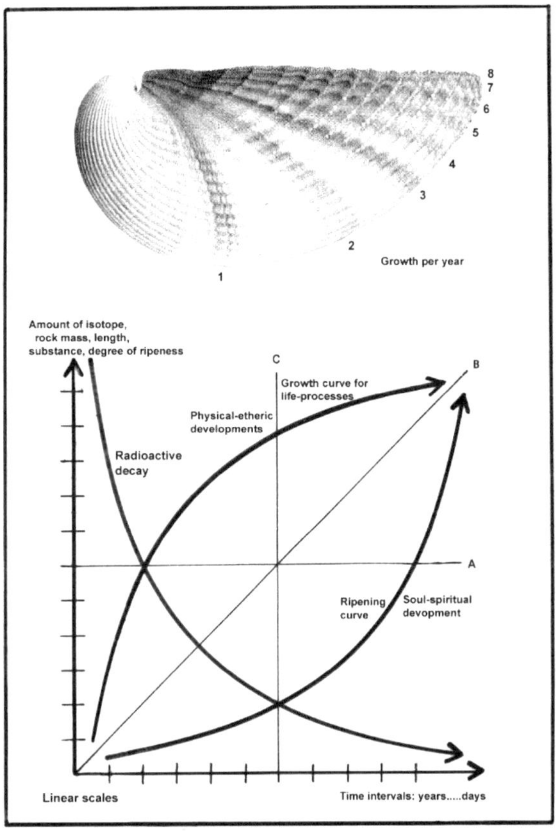

Fig. 82: Above: the yearly growth rings of a Heart Mollusk (Fulvia mutica). Below: The growth curve of the life process and the curve of soul-spiritual ripening in relation to the half-life decay curve of radioactive isotopes. The curves show their polarities and their similarities in that they are mirror images of one another: the growth curve and the decay curve reflected about the line A parallel to the X axis, the soul-spiritual ripening curve and the physical-etheric growth curve about line B, and the soul-spiritual ripening curve and the radioactive decay curve about line C. (For their considerations and corrections regarding questions of dating dealt with in this section, I thank Luc Vandecasteele, Gent, Belgium, and Heiner Frei, St. Gallen, Switzerland).

As an example of the Earth's declining life-processes, one could consider the average thickness of the principal rock layers – the peridotitic, basaltic, granitic, etc. These decrease exponentially! The thicknesses vary within wide bounds, depending on local conditions and depositional areas. For the Earth's crust, one gives the following ranges (from below upwards) peridotite layer: hundreds of kilometers; basalt and granite layers: roughly ten kilometers each; crystalline schists: thousands of meters; sedimentary rocks: hundreds of meters; unconsolidated sediments, tens of meters. The radiometric time for each period gives a rough expression of the thicknesses: Archean 2,100 million years, Proterozoic 1,958, Paleozoic 291, Mesozoic 185, Cenozoic 65. The exponentially-decreasing thicknesses would, under the assumption of inorganic processes, correspond to a shortening of the time required for each step going towards the present; under the assumption of life-processes, they could indicate reduced deposition in roughly equal time-periods, corresponding to a decline of these life-processes, and ultimately point to aging and dying-off of the Earth – comparable to life-processes today in an individual organism.

The second curve shows the rate of soul-spiritual ripening, as the number of developmental steps per time interval, for example, learning. The development of the soul-spiritual in a person or cultural epoch proceeds slowly and later becomes increasingly intensive. Seen spiritually, one achieves ever more in a shorter time. Here one could bear in mind what Rudolf Steiner described as the backward-running occult-astral time stream, which works into the present from out of the future, meeting what comes from the past. Here the cause of an event lies not in the past but in the future. [148] This subject was spoken of when dealing with the double evolutionary spiral.

What connections are there between the three exponential curves? What are the relationships within this "series of interdependent facts concerning their contents"? The growth curve and the decay curve have an essentially similar character; they run as mirror reflections of one another along line A: they begin with numerous steps that decrease steadily over time. The radioactive decay decreases exponentially since with each half-life time interval, half of the remaining substance decays. Organic growth also decreases exponentially. However, with the growth curve, it is not the quantity of substance added which is plotted against time, but rather the total length, weight, size, etc. If the added length (or weight) is plotted on the y-axis of the growth-curve, the curve would be identical to the decay curve.

The ripening curve shows an acceleration of development going forward into the future. The growth curve, on the contrary, a slowing down. With this, the etheric body's life-processes are shown to be polar to the ripening in the astral body. They reflect one another along line B. Suppose one pictures this soul-spiritual curve as running in the opposite direction. In that case, it will represent the aforementioned backward-running time stream. The ripening curve is the polar opposite of the decay curve. Ripening occurs slowly at first and later achieves increasingly more steps in the same time interval. The quantity of a decayed substance per time interval is large at first and later smaller with decay. The curves mirror each other. The decay of isotopes proceeds quickly, while later, there remains ever less of the not-yet-decayed substance.

Life-processes are not directly measurable utilizing radioactive decay. A radiometrically-measured age can determine the point in time when a system becomes physically-chemically closed. Interpreted according to present-day conditions, this means that the determination reaches back to the point where the rock cooled to the temperature where no further gaseous isotopes could escape (Curie point). At this point, the process solidifies to a fixed picture. A series of events, a process in time, passes over into duration. Radiometric dating thus shows the moment where processes cease. The decay curve would then be better placed in Fig. 82 at the end of the curve for life-processes. Under the assumption that the Earth is to be viewed as a living being, the decay curve begins only where this latter has partially died. The decay curve cannot say anything directly about the time-relationships of the life-processes themselves.

An additional problem presents itself – the type of closed system necessary for radiometric dating is only reached by crystalline rocks and not by sediments. However, the boundaries of geological time-periods are defined using fossils in sediments. Decade-long comparisons were necessary to bring the crystalline rocks' radiometric dates into a relationship with the succession of geological formations. In the meantime, a scale had been agreed on internationally, as can be found in geological literature, in museums, etc. The boundaries of further subdivisions can often only be roughly approximated.

Radiometric dating of organic substances has required strong revisions in recent years. Carbon isotopes with short half-lives made time calculations possible into historical times when connected with dendrochronology, such as wood comparisons from Egyptian gravesites. These dates have often had to be significantly adjusted so that the values determined between 1960 and 1980 had to be shortened by up to half their value! [149] The C_{14}-content of the atmosphere, on which the method is based, is subject to strong variations. Carbon isotopes decay at different rates, dependent on time and place. The isotope exchange with the ocean's surface is highly complex and cannot be predicted exactly. The further decay of C_{14} into N_{14} is not measurable since the nitrogen escapes, etc. It is also reported that large spherical carbon molecules have a modified radioactive decay rate. [150]

Another fundamental problem is that decay-rate values have only been measured during the past century. This span is only a short moment when compared to the length of the Earth's history. From the asymptotically running end of an exponential curve, the climb to the beginning must be calculated. Is isotope decay as constant with life as with the inorganic? Did radioactive decay have a beginning? Could the half-life values have gone through development before reaching their present condition? Physics today considers it empirically certain that the physical constants have not altered throughout the Earth's history. However, this point of view can be as little proved as the opposite. These questions can only be clarified when broader concepts are considered.

One of these would be the question: from what time forward can one apply the lawfulness of inorganic nature to the Earth? Consider that in the earth's early history, other physical conditions prevailed than today. Rudolf Steiner explained the differing physical relationships of that time as being due to the states of matter not being separated from etheric forces as they are today. "Physics is not eternal. It no longer has validity for entirely different types of existences. An existence where gas is inwardly luminous (as in the Hyperborean epoch) is an altogether different one from where gas and light are relatively independent [...]. Because in the moment when we reach the sphere where gravity ends, and everything streams outward, in that moment our entire physical picture of the world ends [...] It is just as nonsensical to think that these physical laws are applicable in the far-distant past as well as after a certain point in the future. That is the madness of the Kant-Laplace theory, that one believes what has been abstracted from the present-day physical phenomena can be applied backward as far as one wishes." [151]

By questioning modern conceptions in this way, there is, of course, the danger of opening the door to haphazard concepts regarding earlier Earth history. It can also be taken as an incentive to approach the "mutual evolution of Earth and humanity" from the perspective of present-day life processes. For example, Rudolf Steiner compared the early situation of the Earth and gravity as being similar to the embryo suspended in amniotic fluid or the brain in cerebrospinal fluid because the effect of gravity is mostly eliminated. Present-day physics and chemistry [...] is derived primarily from lifeless nature. "[...] but nature was not always that way. It was once different. [...] Where is there the possibility of coming to the original condition of being? The corpse is the final condition of the being [...]. Anthroposophical spiritual science must bring together this corpse-like condition that existed at the beginning as a living condition preserved only within man himself. In earlier epochs of world-evolution, Earth-evolution was also externally perceptible. At one time, there were entirely different processes active in nature, encompassing nature's origin. Today we go about on the corpses of what originally existed at the beginning. But in the lower part of man, these original conditions are preserved. One can find there what existed earlier, even back to the Saturn condition." [152]

One may occasionally encounter the view that Rudolf Steiner said radioactivity began at the time of Christ. This comes from a statement regarding the concrete example of Radium 226, which has a half-life of 1620 years: "Physics knows today that radium that was present upon the surface of the Earth before the time of Christ's birth is now fully decayed into stable daughter products. Radium that is found today had to have been created 140 years after the birth of Christ." [153] In another lecture, [154] Rudolf Steiner commented that there was no radioactivity in early times. It began a few thousand years ago when atoms began to break up. Before that time, they were becoming more solidified. For Rudolf Steiner, such statements stemmed from concrete situations he was observing, generally about human beings. This instance should not be understood as an abstract or generally valid indication but instead as applying to the condition of substances in the middle of the Atlantean epoch.

Rudolf Steiner on Absolute Time Indications

When Rudolf Steiner spoke of historical Earth events, he sometimes referred to millions of years. These may oppose the time-concept discussed above, which considers a series of events and facts dependent on one another through their content. But when one pays attention to Rudolf Steiner's manner of expression, it shows how he applied this concept of time. He wrote that it is difficult to give numbers and difficult to understand, as one must know the methods that spiritual science uses. [155] In most places, he speaks of millions of years in the sense of unimaginably long periods during which specific evolutionary steps took place. For Rudolf Steiner, it is a qualitative characterizing expression for duration. He describes how with the Gnostics there was still a feeling of the liveliness of time. [156] One did not speak of an abstract timeline running from the past into the future, but instead of beings; he named these creative spirits "Eons." Time was represented in the imagination. In the sense of a hierarchical ordering of spiritual beings, each hierarchy passed the creative impulse on to the one following, bringing it a stage closer to actualization. In other places, Rudolf Steiner referred to the great epochs of evolution as Eons, which can be symbols for what we experience during the year. [157] He spoke of millions of years as Eons, but not in the sense of numbers of years, which would correspond to calculations for the Earth's history using radioactive half-lives.

Rudolf Steiner spoke of "millions of years" primarily before 1910. After this, he began to take a new approach to the question of time. He said that one could get little feeling for the matter from an explanation that divine-spiritual powers had worked for millions of years on the human body. [158] Between 1917 and 1924, he increasingly turned against calculating millions of years into the past or the future. He said it would be like taking the changes during human adult life and extrapolating them for hundreds of years. Geological calculations are quite correct and completely logical, but they do not correspond to reality because millions of years ago, the Earth was not the way it is today. It's as if I could calculate my physical form hundreds of years ago based on relationships in the present. "Long time-periods are for present-day hypotheses just that, something purely hypothetical. Time can only begin to have meaning in an explanation when one derives time from other, concrete presuppositions, when one can form, to some extent, the before and after out of the concrete. Such is not so when one simply presents a developmental current and then brings in time as something external. Rather, time forms itself out of the former." [159] Rudolf Steiner is looking here at relative time-sequences, not absolute ones. As an example, he mentions later in the lecture that the head-organism is the oldest member of the human being. The remainder is younger. This time difference is an essential evolutionary law dealt with in Part I of *Mutual Evolution of Earth and Humanity* by this author [160] when discussing animal evolution. With this example, we have two sequences of sense-perceptible facts dependent on each other: the one being the year, the other being human evolution.

Patterns of Geological Events

In the second column of Fig. 81, the attempt is made to show a correlation with the third column's years according to the precession of the equinoxes using a hyperbolic curve for the radiometric dates. Through this, it becomes possible to compare the historical events of the Earth with human cultural epochs. The human cultural epochs relate to the *Platonic Year*, which is structured by the spring equinox's precession through the zodiac. It is not primarily a question of whether the timescale itself is valid (which would be a Newtonian conception of time); instead, it shows how Rudolf Steiner's time concept makes it possible to arrive at comparable sequences of events that have factual content. The presentation attempts to place the various series of events into an integrated picture.

Dieter Kötter [161] addresses "Logarithmic Measurement of Time" as a special case of a projective presentation for radiometrically derived data, whereby he does not attempt to reinvent an independent "Time Vessel." For Kötter, it is a matter of finding relationships that can lead to an organic measurement of time. The mathematical derivation of growth curves indicates a recession of life processes appropriate for the living Earth. Kötter infers that various hyperbolic curves are indicated for geologic ages, depending on the time-periods for which they are used. He maintains that the mathematically necessary thresholds are understandable as phases of change, where a different individual timescale for the sequence of events becomes meaningful. A logarithmic scale with a base of 3.18 is used between the limiting values of 3.8 billion years and 3,000 years BC to encompass the entire history of the Earth with one timescale and without such thresholds. Significant milestones provide the fixed points required for determining the logarithmic scale in the fossil record – the radiometric time of the Precambrian/Cambrian boundary at 575 million years, the Cretaceous/Tertiary boundary at 65 million years, and the high-point of the Ice Age around 500,000 years.

An "absolute" time scale in years based on radiometric dating is so problematic that it cannot be considered a standard for Earth's evolution as a living being if we look at all the above aspects together. Isotope decay cannot say anything about the time conditions of life processes since it only begins below the Curie point. Also, there are the questions of inconsistent half-lives and their beginning, the uncertainties using the example of carbon isotopes, and the emptiness of the abstract millions of years. On the other hand, Rudolf Steiner's concept of time makes it possible to arrive at a concrete relationship of the time conditions by distinguishing four main areas in which the geological facts form dependent sequences of events according to their subject matter. In contrast to the millions of years, they have continuous subject matter. Therefore, they have practically always been used by researchers as relative reference sequences. The first sequence is that of the annual rhythms. According to the past, they form a meaningful measure of historical events up to about 3,000 BC. This time is when the individual history of humans and the historical legend begins with the Babylonian-Egyptian cultures. It is not that the year before no longer exists, but it is increasingly becoming an abstract mathematical quantity. The calculations are correct; only the calculated object has disappeared. Calculating years for the history of the Earth is like the attempt of an older human who wants to remember their whole life by the second or minute. I can undoubtedly calculate when I had an experience 2,763 days ago, but the number does not mean anything to me because I have forgotten the sequence of daily intermediate events. I have to move on to series of events with larger temporal steps.

Before that, the Platonic Year offers a possibility to compare the developmental steps during early human cultures with the rhythms of the Ice Age, roughly back to the middle of the Atlantean period. For a cultural epoch, 2,160 years were set as a twelfth of the Platonic Year and calculated back from 747 BC. For the expansion of the constellations, the values of Gunther Wachsmuth [162] were used. The events of the Earth's evolution are also radiometrically dated for this period. Both sequences could be compared if the second half of Atlantis were concurrent with the Ice Age (Fig. 81). The cosmic relations between Sun and Moon correspond to those of today so that the year can still be calculated as an abstract unit. However, to go back even further with the concept of the year would be as unrealistic as the calculation of millions of years in radiometric dating.

For the epochs before the Platonic Year, i.e., the Tertiary period, the Mesozoic, and Paleozoic eras, the animal kingdom's development is the most densely populated event sequence. It forms a dependent series with the manifold rhythms and characteristic rates of the rock processes, the Earth's crustal movements, the climate zones, and plants' evolution back to the Cambrian period. For the chronological classification of the fossil-free crystalline rocks of the Precambrian, the calculated numbers from the decay series of the isotopes have proven to be a relative reference value. Radiometric dating can be used to link rock complexes into sequences that can be compared worldwide or with other event sequences, such as in the correlation of oceanic basalts or granites in mountain massifs. This method produces a coherent overall picture, which shows that the procedure is readily applicable in this sense, corresponding to the fossil sequences in the sediments. The "absolute age" is mainly spoken of by popular science and museums. In research, it is more realistic to use the term radiometric dating.

Ther determination of a rock's age requires it to be placed in an interrelated series of events running parallel to its history, which are arranged in order of occurrence to bridge the period of its formation to the present. Four different groups of content-dependent factual sequences make it possible to compare historical and geological processes in terms of content:

1. The determination of the quantity of an isotope for relative radiometric dating in fossil-free crystalline rocks, their tectonics, hardening, and crystallization;

2. The evolution of animals and plants in the rocks of Earth's recent history;

3. The Platonic Year with the Ice Age and prehistoric cultural epochs; and,

4. The year concerning traditional historical developments from about 3,000 B.C. onwards.

15. The Earth was a Living Being

The Phenomena Prove that Rocks Originate with the Earth's Life Sphere

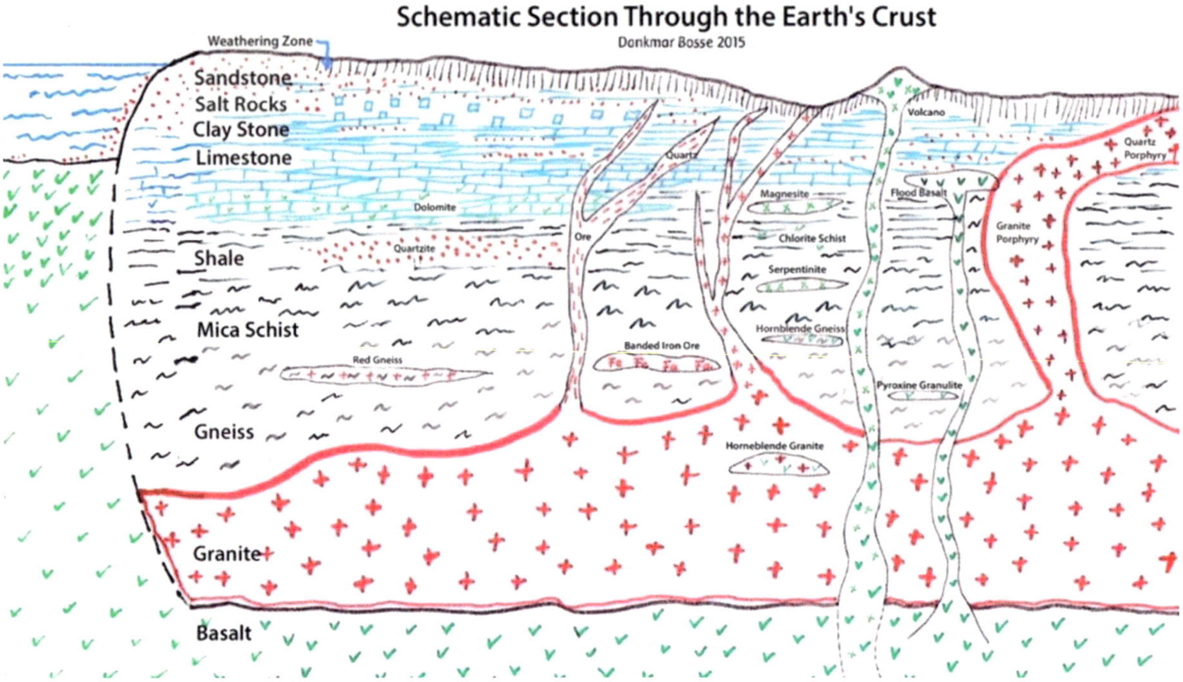

Fig. 83: A section through the Earth's crust.

Fig. 83 shows a section through the Earth's crust; the continents are on the right, the oceans on the left. One can see how the Earth's continental crust begins with a granite layer, which is missing over the other two-thirds of the planet where the basalt and olivine layers lie directly under the oceans. The oceans spread over two-thirds of the Earth's surface; the continents take up only one-third. The continents originally formed together as one large landform; their granite layer rests on the lower basalt shell and is about 10,000 meters thick. About five kilometers of gneiss come next on top of the granite, which turns into shist and shale towards the top. These form the *continents' bedrock*, above which limestone, clay stone, and salt rocks are deposited in the *overlying strata.* The sandstones are added from the erosion of bedrock. Above this are the sands and clays of the *unconsolidated rock*. This very general sequence, as shown schematically in Fig. 83, is uniquely formed in each area of the various continents, depending to what extent these layers were lifted and removed by mountain building. In the entire cosmos, the rock sequence's uniqueness from granite to sediments is only known to exist on the Earth!

The Earth is entirely different under the ocean waters, where the basalt (or gabbro) is the ground under a thin layer of loose sediments. It also passes under the continents with a thickness of about 30 kilometers and forms the Earth's solid crust. Only through volcanism (Fig. 83, right) does the basalt reach the surface, where it flows out as lava or as a flood blanket because the impulse of volcanism has an effect that goes deeper and pushes the basalt upwards through cracks and fractures in the Earth's crust. Even deeper in the upper mantle (below the basalt layer), there is an olivine (or peridotite) layer. An iron-nickel zone can be assumed In the lower mantle, which causes the Earth's magnetism. But not in the Earth's core, concerning which no statements can be made. There are only different hypotheses. The Earth has these mineral masses beneath the oceans and continents in common with the solid planetary bodies of Mars, Venus, Mercury, Moon, and the asteroid (planetoid) belt between Jupiter and Mars.

How did all of this come about? The natural sciences always describe several models in popular writings to answer this question; however, they cannot answer them in the scientific geological literature. For example, it is assumed that granite was formed from sediments through melting, and then sediments are formed again by weathering and erosion. With this *rock cycle,* one question remains open: from where does the primary granite originate? This problem is described in my book *The Mutual Evolution of Earth and Humanity*. [163] It cannot be extrapolated from the basalts because they contain less silica bound in its minerals and because many other minerals originating in the granite are missing. Carbon appears in the gneisses above the granite layer, which only on Earth occurs in huge quantities. It is first finely distributed as graphite in the black clay shales, concentrated in the bituminous coals, and finally in the brown coals. It is generally assumed that the coals are made of peat. Considering that one meter of fresh peat becomes about 20 cm of brown coal or 2 cm of bituminous coal speaks against this assumption. The largest brown coal seam contains 320 m of coal, which would correspond to 1,600 m of peat; the largest bituminous coal seam is 200 m thick in Fushun, China, which would require 10,000 m of peat! How could this come about from underwater decomposition – in shallow depressions supporting plant life that would also have had to extend over several hundred kilometers? In general, it is necessary to become aware of the Earth's size when considering geological processes. Dimensions, quantities, spaces, and times that are difficult to imagine must be scrutinized!

The first limestones appear simultaneously with carbon, first as fine crystalline marble, then from bacterial metabolic processes formed by the lower animals, later from their shells and skeletons. Chemically the limestones consist of carbon, calcium, and oxygen. Calcium is already contained in the feldspars of the basalts – but the carbon is not. How can these simultaneous formative processes of plants and animals, which were new and only present during the Earth's early development, be understood? And from where do the 363,500,000 cubic kilometers (!) of ocean water originate? Or the air with its 21 % oxygen and 78 % nitrogen, instead of the 3 % on Mars and Venus, where the atmosphere contains 95 % carbon dioxide, whereas, on the Earth, it contains only 0.03 %. From where do all these differences that exist between the Earth and the cosmos come?

Many fundamental questions in geology remain open for the geosciences to explain the Earth's origin, including all that uniquely exists on the Earth, from the rocks up to the human being. Materialistic research describes many phenomena – especially their usefulness for humanity. That is its task, and it fulfills it very well! But this materialistic research stops at its limits of knowledge, where it offers to the general public sophisticated models that often contradict each other: Our solar system was formed 4.5

billion years ago by the contraction of a cosmic dust and gas cloud. All substances on and in the Earth were created by combining atoms and molecules. Life, and thus also humans, arise from chemical-physical processes that were the product of chance. All of this is taken for granted and is no longer mentioned in the models' description.

This view is based on the method of geological actualism. It assumes that the dead Earth's now dead processes can be extrapolated into the Earth's past. Suppose the phenomena of rock formations are examined more closely. In that case, they show lawfulness and patterns that can be observed in current life processes and from which the inorganic continuously emerge. It leads directly to the idea of a life sphere. In this way, a comprehensive picture of the mutual evolution of Earth and humanity can appear. The phenomena of natural science and Rudolf Steiner's representations illuminate each other and go beyond the limits of knowledge and materialistic models.

An originally living Earth body can, as an initial approximation, be compared with metabolic processes of the lowest bacterial-plant-like organisms, which can still be observed today. However, one would have to imagine a uniform colloidal-bacterial life process in a hot-fluidic state, airier above and becoming denser below. From this, the rock-forming substances have successively precipitated onto a nascent Earth surface. The details of this becoming can be read from these sequences. Thus we can recognize something of their essence, as if from a biography.

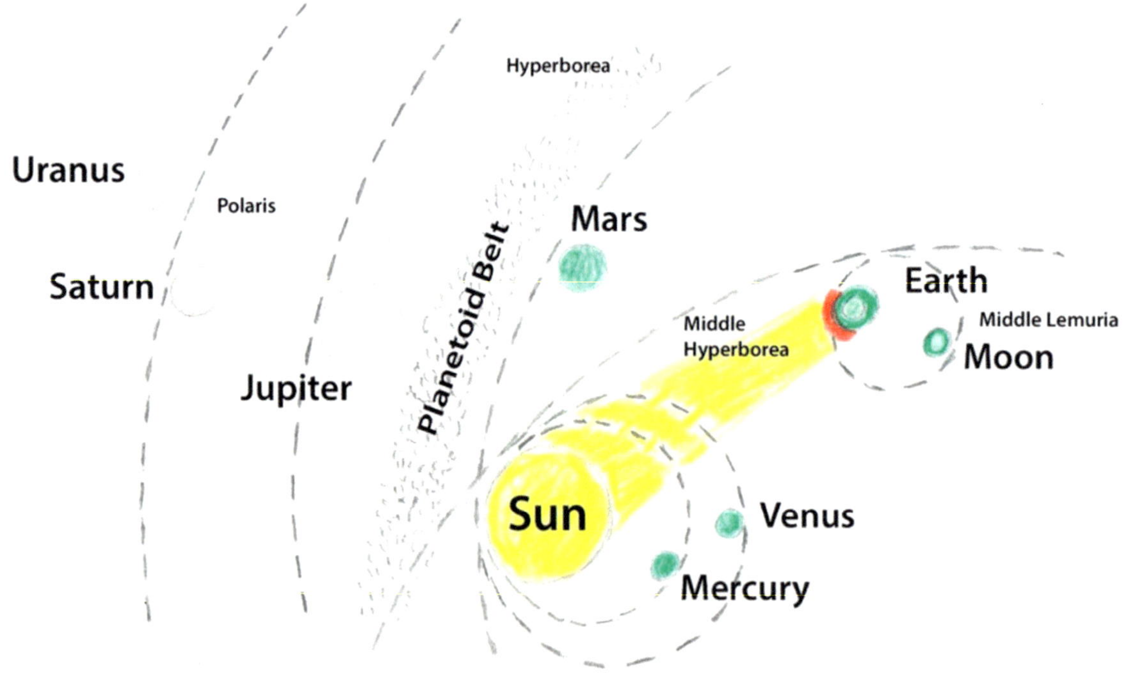

Fig. 84: The separation of the planetary bodies through the contraction of their spheres of activity. Rudolf Steiner sketched only the circles on the drawing and wrote the names that designate the planets. I took the liberty of adding the boundaries of the spheres, the planetoid belt, and the ages. The sizes and distances cannot be represented to scale in a drawing.

Fig. 84 shows the beginning of this development. In a lecture [164] related to this subject, Rudolf Steiner drew only the planetary orbits. I added the planets, the planetoid belt, the names, the basaltic shell (green), and the Sun radiance. From the Polarian time on, the planetary bodies gradually contracted. The planets were left behind as materially visible inclusions on the outer edge of their orbits. The same is true for the Earth body, which was still in a hot-airy-fluidic state in the early Lemurian Age. It did not yet rotate and was illuminated on one side by the Sun, bringing about the life processes. Thus, above the still fluid basalt layer, the granite layer was formed in the great continental area that only in the middle of Lemuria, at the end of the Paleozoic era, finally broke apart when the concentration of the Moon basalts concluded. [165]

The subsequent development paths for granitic minerals and their forms correspond to the basic laws applicable to the evolving natural kingdoms. These include diversity, individualization of mineral substances, and the laws of metamorphosis relevant to evolution and devolution. For example, the numerous lens-shaped rock and mineral bodies point to colloidal processes. They testify to the gradual separation of plant and animal life reflected in the graphite and crystalline limestone. These developments pass over into the shales and limestones in which life processes work toward rock formation, which can be clearly observed today. These primary substance processes are ultimately isolated from humanity's evolution, as described in *An Outline of Occult Science* [166] by Rudolf Steiner. Next, in the Lemurian period, oil, natural gas, and gas hydrates were exuded from the life sphere and mostly came about in huge quantities at the end of the Mesozoic era. Although biochemically different, they are derived from lower animal-plant remains. But why didn't they develop in this way in other geological periods? These ideas make it understandable that Rudolf Steiner repeatedly speaks of an albuminous atmosphere that decayed into the separation of hydrocarbons. The minute sulfur content of the protein is also shown, for example, by the coloring of weathered limestones. Then appears the light yellow color of pyrite, an iron-sulfur mineral.

All that remained at the end of the Earth's evolution was a swirling air-water sphere, which can be imagined if we think of today's ocean vortices and air cyclones combined. During the Ice Age, the last precipitation of water took place. This event corresponds with Old Atlantis, about which legends still speak. The Earth's landscapes were formed by these enormous precipitations described in great detail in my Mutual Evolution book. [167]

When the coagulated, crystallized images of the worldwide landscapes, rocks, and minerals are "read" and examined in detail, new access to the mineral kingdom becomes possible. Unfortunately, the real development of the Earth and its natural kingdoms to the human being remains closed to today‘s education and popular science, which leads to all the damage that today is destroying our life foundations. In contrast, we can learn through the Goethean method to perceive the spiritual side of the sense world imaginatively in the physical world's phenomena. There is no gulf between the two opposite worlds over which a "bridge" must be built. The facts themselves lead on a continuous path to the imaginatively experienced idea of the Earth's life sphere, which is the key to a new understanding of its development.

Thus Rudolf Steiner says in a public lecture: "Everywhere, as you will see, spiritual science links up with natural science in such a way that the latter on the pure basis of facts provides ample proof of what spiritual science as such has to offer. Never are there any grounds for contradictions between spiritual science and true natural science. [...] If you ignore those popular works which declare this or that to be a "scientific fact" and go to the sources, then you will find, especially in the field of geology, that geologists everywhere get to a certain point — and then put a question-mark. Arriving at those question-marks, one can take them as a starting point for spiritual scientific research. [168]

16. The Contradiction Between the Atomic Hypothesis and the Phenomena

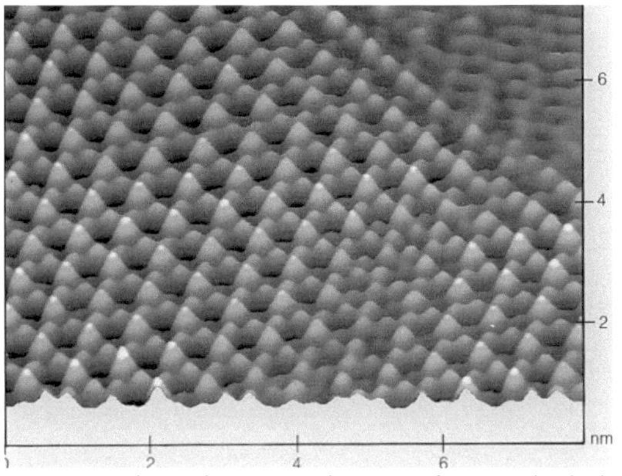

Fig. 85: A three-dimensional image of an octahedral surface of iodine on platinum under a copper-containing electrolyte. (Brochure from Digital Instruments, 1992.)

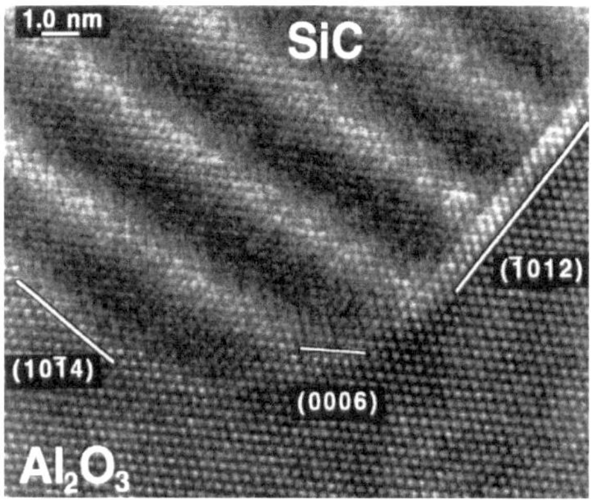

Fig. 86: The boundary between silicon carbide and aluminum oxide with its different lattice structure.

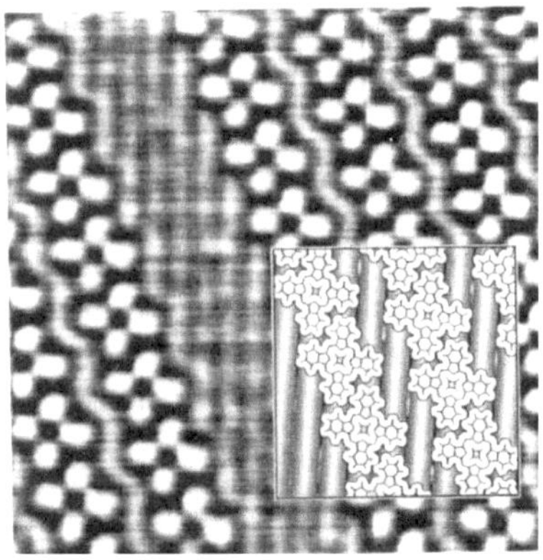

Fig. 87: Copper phthalocyanine on molybdenum-selenium with an insert showing its molecular structure, 10x10 nm.

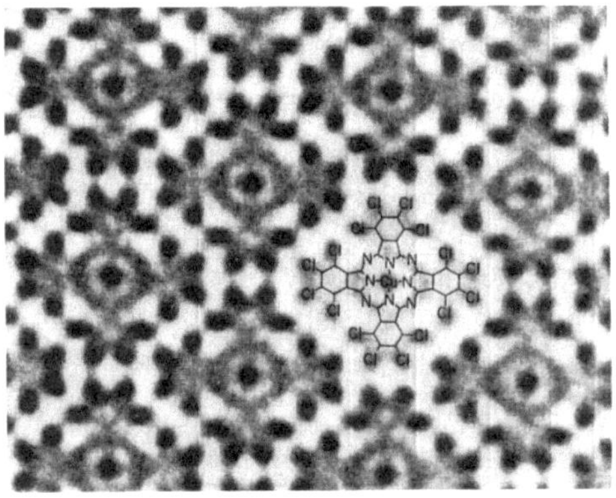

Fig. 88: Copper phthalocyanine on alkali halide with an insert showing its molecular structure: a copper atom surrounded by 16 chlorine atoms.

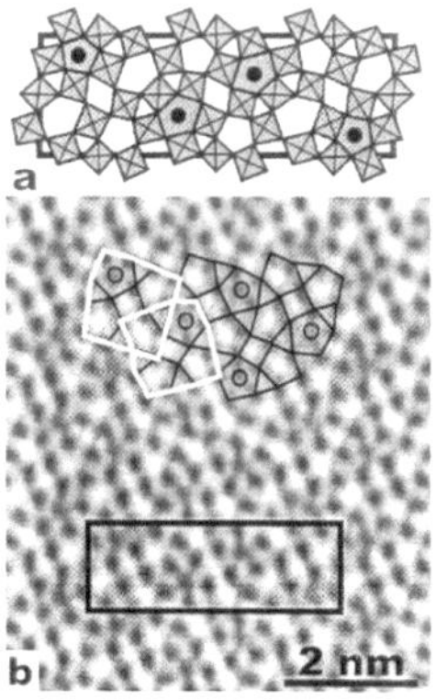

Fig. 89: A structural image of niobium-tungsten oxide in two different valences, with an insert showing its molecular structure. (Sayagues, M. (2001) in Micron 32, 5, p. 465, New York.)

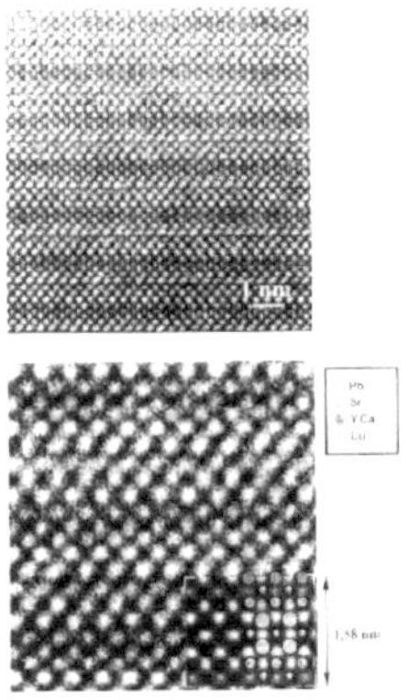

Fig. 90: Lead strontium copper oxide shown above. Below it is magnified, with four cations highlighted on the lower right and identified in the upper right box. (Alexander, H. (1997): Physics. Electrical microscopy).

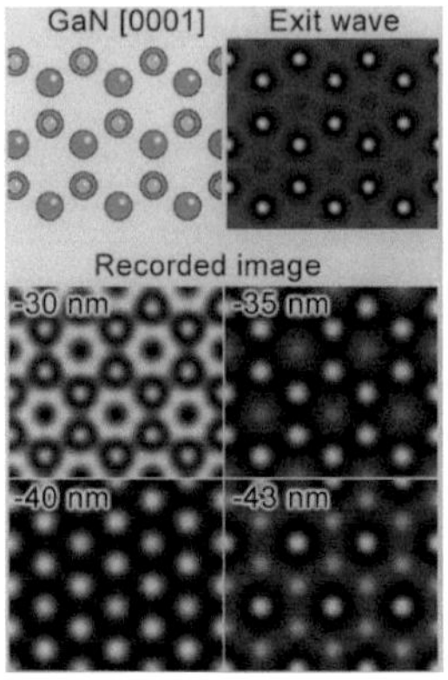

Fig. 91: Top: gallium nitrate. Top right: exit wave; Lower Left: a structural model for it; Bottom: different imaging magnifications. (Til Bartel (2006), www.wikipedia.org).

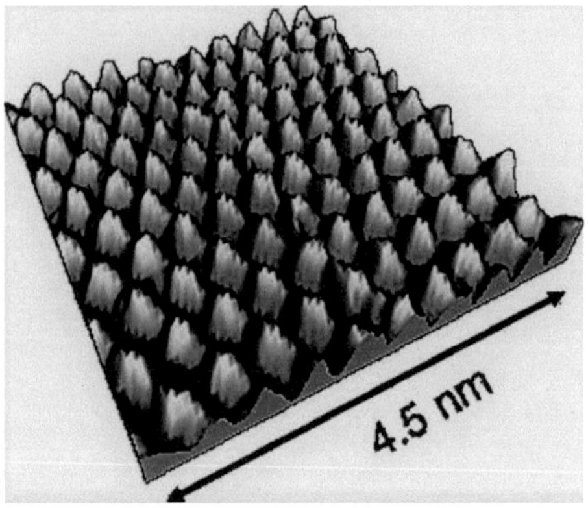

Fig. 92: The surface of table salt recorded with an atomic force microscope in non-contact mode. (Shielding irons, A. Weiner, 4.1.2007, University of Münster, www.wikipedia.org)

Modern electron and atomic force microscopes provide much more detailed images showing the inner structures of substances than those obtained with X-rays. It is not the substances that are important in these images, but only the structures and their interpretation, which are described below. More delicate structures can be depicted with ultrashort wavelengths. Ultrashort wave electron beams make extremely high-resolution possible, which leads to the extreme limit of what can be depicted. The human eye reaches a maximum resolution of up to 0.2 mm, a modern light microscope up to 0.00038 mm (= 380 nm; 1 nanometer [nm] is 1/1,000,000 mm). The highest currently achievable resolution goes down to 0.1 nm (= 1 Angstrom = 10^{-10} m), and modern atomic force microscopes achieve this.

The complicated technology is described as follows (in excerpts, shortened from a Wikipedia article): "The atomic force microscope is an essential tool in surface chemistry and is used for the mechanical scanning of surfaces and the measurement of atomic forces at the nanometer scale. A measuring tip, located on an elastically flexible lever arm, is guided as a measuring probe a short distance above the sample surface. A piezoelectric scanner moves either the tip over the sample or the sample under the fixed tip. An atomic-scale fine point can be achieved by using a single carbon monoxide molecule as the tip. The lever arm's deflections caused by forces between the sample and tip are measured at high resolution, usually by directing a laser beam at the tip and then collecting the reflected beam with a photodetector. An image of the sample surface can be generated by recording the deflections or forces point-by-point, just like a digital photo. Each image point then represents a specific measured physical or chemical quantity. The non-contact mode is usually used in vacuum or ultra-high vacuum, where it achieves the highest resolution compared to the other operating modes of the atomic force microscope.

In contrast to the high-resolution scanning tunneling microscope, which achieves atomic resolution on electrically conductive samples, even individual atoms and molecules on electrically insulated surfaces can be imaged. Thus, the atomic force microscope and the scanning tunneling microscope have the highest resolution of all microscopic techniques. The reflected signals are made visible on the screen as circular areas of varying degrees of light and dark and are displayed as two- or three-dimensional surface images."

Through the technical means of light microscopy, X-rays, and electron beams, images of ever more delicate structures can be obtained up to the limit reached by atomic force microscopy. This limit is where the possibility of resolution and representation ends, thus the endpoint for *photographing atoms*. No qualities such as surfaces, colors, or chemical properties can be perceived here as is possible at the macroscopic scale. The images show regularly repeating structural units, which are interpreted as intersections in a three-dimensional *lattice of atoms and molecules* forming *networks* of parallel planes. The structural images shown as inserts in Figures 87, 88, and 89 indicate how molecules are imagined. The limit of knowledge lies between the surface of matter excited by laser beams and the models derived from the phenomena! When we refer to atoms or elementary particles, we at first imagine particles of matter. However, in the pictures, we do not look directly at the structures of matter but at surfaces where discrete energy impulses move around and interact in different states. Suppose we want to characterize this complex, widely branched, and complicated research field in a simplified way. In that case, we could better say: These energy impulses are called *atoms*. They consist of a nucleus orbited by electrons at almost the speed of light, in orbits that are bowl-shaped and located at specific distances from the nucleus. They are pictured rapidly moving in a vacuum through *electrostatic forces* with their charges spaced and kept in balance by electromagnetic forces of attraction and repulsion or as standing waves. The individual components of the atom and several hundred other types of *particles* are only interactions of electromagnetic and other forces that can decay into ever more *energy quanta*, or they are interpreted in terms of their possible transformations. Also, sunlight or electric light is invisible; light only makes things visible. It is understood to consist of a quantum of energy (photons): *the photon is the field quantum or the electromagnetic force's binding particle*. In today's materialistic science, the atom and molecule are taken as a self-evident reality, as if there could be nothing else, and from there, the causes of all sensory phenomena are explained and justified.

For example, Prof. H. J. Rösler, in his mineralogy textbook, begins the description of crystalline forms with the sentence: "Since it is known that all matter is made up of atoms (or ions and molecules), one can conclude that these small building blocks are not randomly distributed, but rather cause this phenomenon by regular arrangement." [169] And a little further on, it says: "Since the form of a crystalline mineral is determined by its atomic structure, its crystalline structure, there is a correspondence between the ‘symmetry of the atoms’ and the symmetry in the real crystal.” This statement leads to the fundamental question: Where do the crystalline forms come from, which are so regular that they can be organized into exactly seven systems, 32-point groups, and 230 space groups? Why are the angles between the surfaces 45, 90, or 120 degrees, with deviations of at most plus or minus exactly one degree? Or is it because the *atoms' symmetry* is derived from the actual crystals' shape and symmetry? A Viennese professor of mineralogy writes: “The implied principles of symmetry, which are so clearly expressed in the world of crystal forms, present us with the great riddle of how the formative forces expressed here is achieved. Since we regard crystals as homogeneous bodies – which, according to common terminology, should be evenly and completely filled with the substance in question – it seems incomprehensible that the crystal surrounds itself with flat surfaces. And should we not ask why it exhibits those puzzling symmetric peculiarities? From ordinary homogeneous matter, we would expect much more spherical or grape- or kidney-shaped forms or even completely irregular forms.” [170]

These "molecular" structures are "infinitely" extended in all three dimensions relative to the crystals' actual size, considering the atomic force microscopic images' enormous magnification. Why do the *regular geometric forms* have a finite harmonic shape instead of a random, unshaped heap of primary unit particles? Only their lines correspond to the electromagnetically generated images. In contrast, Figs. 85, 86, 87, and 90 illustrate internal boundaries between different crystalline regions. It is astonishing how strictly linear these boundaries are, running partly parallel to the electromagnetic structures in Fig. 87 and independently in Fig. 86. These internal boundaries are attributed to *energy conditions* but do not say much more. No less than eight (!) different models are listed as an explanation for this. The question of qualities inherent in all things also remains open, such as different colors, tones, hardness or softness, the developments and metamorphoses of plants, animals, and also the ordered processes within the human being. Are they also somehow caused by the non-qualitative electromagnetic forces of atoms and molecules?

The outer crystal form is created when the substance changes from the fluid state to the solid state. It starts from a crystallization point that is already solid. A snow crystal, for example, starts to crystallize on a grain of dust. Crystallography leaves open the question of the very first appearance of the form and describes it as a random meeting of ions and atoms, which then form crystal lattices to which other lattices are attached. The beginning is called a *node* in association with the plant world. A living being continues to grow by internal cell division; in a mineral, however, all other substance is taken up from the outside. People often say a crystal grows, but this is wrong because to accumulate from the outside is the opposite of growing by internal forces! A child does not grow by accumulating food on the outside! Growth is controlled from within. On the other hand, a crystal becomes more extensive as long as its substance is present in the environment. There are rock crystals from one millimeter to more than two meters in size!

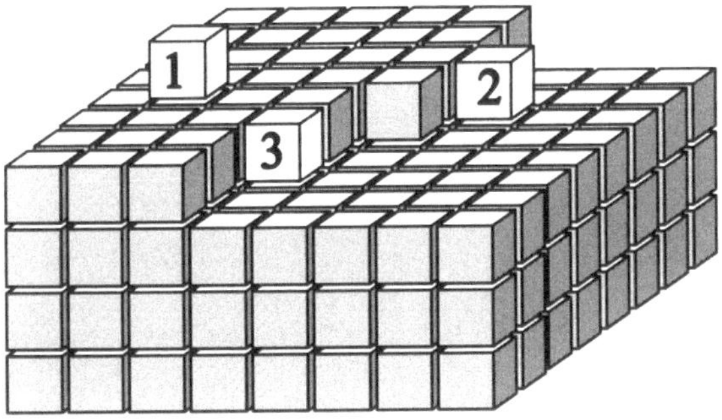

Fig. 93: Shows the model presentation according to Kossel and Stranski for the attachment of molecular lattice building blocks during the formation of crystal shapes. (Rustemeyer, P. (2003), p. 48)

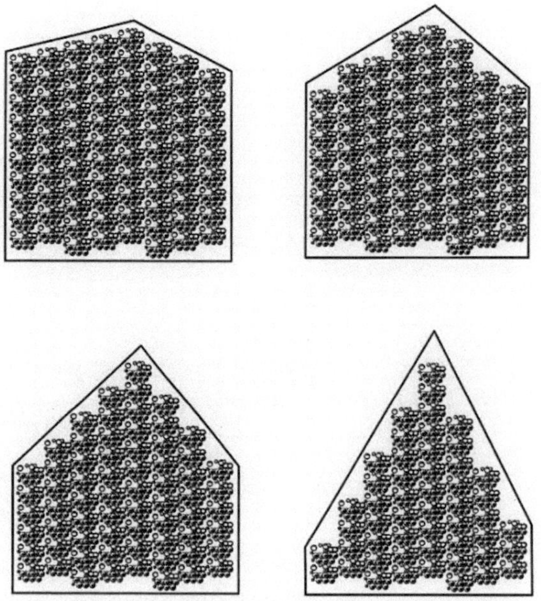

Fig. 94: Shows how four images of different outer surfaces of tourmaline crystals can be easily understood by adding molecular lattice building blocks arranged in a staggered fashion. (Rustemeyer, P. (2003), p. 56.).

Paul Rustemeyer writes about Fig. 93: “At the surface of the crystal (1) one attracting force is active; at an edge (2) two attracting forces are active; and, in a corner (3) three forces are active. The stability of a system increases with more attracting forces at work. So the ion preferentially moves to the most energy-efficient position (3). The least favorable case is a deposit in the middle of a surface (1). [...] Thus during the growth of an ideal crystal, the partly occupied surfaces are the first to fill, and the result is crystal bodies bounded by planar surfaces.” [171] However, the *attracting forces* cannot be measured directly, but are retroactively determined from the force required to separate the lattice element!

Following these ideas, Fig. 94 shows a molecular lattice model that Rustemeyer uses to illustrate four outer surfaces of the approximately 80 complicated crystal shapes of tourmaline. However, he writes: "Just over 10,000,000 of these silicate rings are arranged in a gigantic stack per centimeter of crystal height.” For 8 or 10 silicate rings, the surface shapes can still be illustrated in this way, but never for 10 million per centimeter of crystal. By comparison, if a lattice were one centimeter in size, the outer surface would only be reached at a distance of 2,000 km. And the tourmaline crystals did not grow to just a centimeter, but 30 cm! Who or what would end the lattices' accumulation, so a regular outer shape was created on all sides? When the Egyptians built the pyramids, a pharaoh took care that the same number of cube-shaped building blocks were placed on all four sides. Otherwise, the work would not result in a pyramid.

Fig. 95: Cerussite, radiant, pointed.

Fig. 96: Bismuthine, grown hanging in a hot solution, with emphasis on the edges.

Fig. 97: A pyrite with offset surfaces.

Fig. 98: A quartz crystal, with emphasis on tip and edge.

According to the models of Kossel and Stranski, a surface must always be filled first, then the outer edges, and finally the corners. What do the phenomena on the surfaces of real crystals show about the accumulated substances? The four pictures above show a selection of the numerous crystals that do not have completely smooth outer surfaces. In contrast to the models, these examples always show that the substance first attaches to the prospective corners, then to the edges, and only at the end to the surfaces, provided fluid mass is still available in the surroundings. Deposition at the corners (Fig. 95), which happens very quickly, results in long, thread-like crystals (*whisker, skeletal or dendritic growth*). The three most extended tips already go toward the corners of the prospective crystal! This action is observed mainly in molten states or solutions which are highly supersaturated or supercooled. Several supplementary models were developed in the literature to *explain* this, which assume cooling and diffusion rates, "*push" supersaturation*. Figures 96, 97, and 98 show numerous smaller crystals in the prospective planes that are not yet filled and are formed in the same direction and shape as the overall crystal. This example also shows how formative forces act from the outside. Science associates this with a *building plan*, a *distinct building principle,* or *architectural idea*. But who is the builder, the architect who implements the building's ideas and then makes it visible?

Fig. 99: Fluorite with offset facets.

The example of a fluorite crystal (Fig. 99) is fascinating. It is cubic on the exterior, but on the surface, one can see edges of octahedrons shifted by 90 degrees. The inside is crystallized octahedrally throughout the crystal! You can easily knock the octahedrons out of this crystal, but not cubes. Octahedral crystals of fluorite are scarce. Various minerals show how their substance is always formed from the corners and from the edges to the surfaces. The entire process of formation can be read from this.

Because it is impossible to transcend these limits, we use associations where electrical phenomena are visualized through soul-spiritual characteristics! For this purpose, a few examples are quoted here, whose associative core words I have underlined. The textbook by Borchardt-Ott [172] states the following about crystal formation from lattice building blocks during cooling: "Matter at the appropriate temperature definitely strives towards the crystalline state." And further: "Due to the lattice-like structure, crystals are often bordered by flat surfaces or, in the ideal case, form regular geometric shapes." Paul Rustemeyer [173] writes about coincidence, which explains the *miracle of the formation* of the first crystal nucleus: "The nucleation occurs spontaneously; a shape is created, and the surfaces shift. It is an unbelievable coincidence; [there must be] a powerful attraction between the silicate rings so that these particles in nature repeatedly join together to form the building plan of the tourmaline crystal. The crystal makes compromises and accommodates crystal building methods inserted as visitors into the differently sized lattice sites. These visitors are considered to cause the ideal *building plan* not to appear and why a distinction must be made between the form and the shape." The reason for the difference is attributed to numerous *lattice defects*, which are listed for example by Hausühl: "The following types of building mistakes should be mentioned: gaps, occupation of interstitial spaces, incorporation of foreign atoms, shifts, mechanical deformations, misaligned structural units, errors in chemical composition, deviations from the homogeneity of cells in isotope mixtures, non-homogeneous distribution of thermal movements, sound waves, various electric and magnetic field quantities."

Despite these numerous possibilities for error, regular external shapes and numerous regular twins and triplets are produced, which are attributed to such lattice defects. Rudolf Rykart begins his excellent monograph on quartz with the words: "Crystals can define themselves through geometric shapes. And they continue to strive to assume a form that is as highly symmetrical as possible." [174] As a final example, Paul Niggli describes the transition from lattice to the crystal as the crystal morphological law. He says "that structures of a certain building principle gradually come together in newer, greater unities without completely losing their individuality. The smallest units discovered or invented and normalized by the atomic physics of our time are made up of atoms of different sizes, although only finite. These atoms come together in molecules and crystals to form higher and greater volumes. Here the symmetry relationships in the narrower sense are active, separate the real or near-real from the conceivable multiplicity. We give shape to this striving for symmetry by carrying out ideal designs."

And further: "From the smaller building blocks, one large, fully connected body is not formed, but many different kinds of atoms are formed. Without completely giving up their individuality, they come together to form various molecules or crystal compounds, all of which have only finite dimensions. However, the idea of the building is often tied to the infinite." [175] These professors' citations point in the right direction; that the causes of the forms and qualities of things lie in the soul-spiritual world behind the sense-perceptible world. Not only is the architect missing who has the idea for the *building plan*, the *building idea*, but there are coincidences in the sense of *dialectical materialism, unrecognized lawfulness*. The question remains: Who or what works in this unknown territory? How do I gain access to knowledge there? This problem characterizes particularly clearly the cognitive limits of natural science to such "phenomena."

A professor of physics also concludes that the overall connection of all formations can only be described as follows: "[...] solids and minerals have many new properties that the individual components of the system do not possess, but which only come about through the interaction of a large number of atoms. An interesting question is then how the individual properties of macroscopic systems develop during the transition from atomic to molecular to clusters to the solid state. [...] On the other hand, tracing the properties of macroscopic systems back to the properties of the individual atoms and their interactions remains a major challenge." [176] Ultimately, it is also the task of applied materialistic natural science to describe phenomena as long as there are still phenomena at these magnifications. This concept is quite sufficient for experimentation and utilization. Materialism is defined in the dictionary of foreign words as the *doctrine that matter is the only real thing in the world and the spirit only its properties and effects*. Nevertheless, this worldview has achieved great technical success, as in nuclear-technology – even when combined with deadly dangers and the unsolvable problems of radioactivity. Nevertheless, Rudolf Steiner emphasizes: "Materialism did not originate because material beings and processes alone can be perceived in external nature, but because the human being had to pass through a developmental stage which led to a consciousness that was at first only capable of seeing material manifestations." [177]

Stefan Baumgartner characterizes the concept of matter in quantum physics in a modern, generally understandable way: "What does an atom look like in quantum physics? It is no longer a small, solid piece of matter that behaves as one would expect from an aggregate of small spheres. From the basic concept of quantum physics, an atom consists of electrons in an electrical force field that emanates from the positively charged atomic nucleus. [...] Several possible solutions describe the electrons' energies and residence probabilities for a basic mathematical formula (say, the Schrödinger equation). It is impossible to predict how these possibilities can be observed experimentally; only probability statements can be made. No statement about the "atom itself" can be made if a positivistic attitude is adopted. An electron is also a quantum object to which no easily imaginable material quality can be attributed. A mass, an electrical charge, a location, and a speed can be assigned to it in its effects. Ontologically it does not possess these properties independently of an experimental interaction. [...] Thus not only is the atom non-existent as a materially imagined object, but it is also invalid as an explanation of the world in the sense of atomism." [178]

Rudolf Steiner points out this contradiction in many places, how in science, underlying properties are taken from the sense world, which cannot be found in matter as such. I have selected some quotations leading to the question about what matter is. In 1884, in Rudolf Steiner's explanations of Goethe's natural scientific writings, he opposed atomism. He points out how every natural phenomenon has sense-perceptible and non-sense-perceptible components; substance is sense-perceptible, forms are supersensible. One attributes "[...] a real content to a non-sense-perceptible entity — as the elements of discontinuous matter (atoms) are supposed to be [...]. Although ever so sublimated, there must be sense qualities added to this concept of the initially quite empty concept of the atom. One person adds impenetrability to the atom, another the exertion of force, another its extension in space, and so on; in short, each ascribes certain characteristics borrowed from the sense world. If one does not do this, one remains in a complete void. Therein lies the semblance, the contradiction. One draws a line through the middle of what is sense-perceptible and declares the one part of being objective and the other to be subjective. [...] If there are atoms, then they are simple parts of matter, with the characteristics of matter, and are not perceptible only because their small size makes them inaccessible to our senses. But with this, there disappears any possibility of seeking anything in the motion of atoms that could be held up as something objective, in contrast to the subjective qualities of sound, color, etc. And the possibility also ceases of seeking anything more, for example, in the connection between motion and the sensation "red" than a connection between two processes that both belong entirely to the sense world. Therefore it was clear to the editor [Rudolf Steiner] that motion of the ether, the position of atoms belong in the same category as sense-impressions themselves. Declaring the latter to be subjective is only the result of dim reflection." [179]

"Of course, the phenomena lead to a certain atomic structure in one field or another. It is only questionable when trying to explain the processes within atoms as a whole process or as the complete facts of the case. [...] So it must somehow always lead phenomenologically to these smallest particles, that one has an atomic structure [by a kind of process] – that it is always necessary to explain this atomic structure out of something else, out of the total facts of the case. [...] We are not opposed to the fact that processes that take place in such a small way, like the ones we think of in the atomistic world, are real as such." [180] And later, Rudolf Steiner says: "[...] When you look into the outer world, you realize that it is complete nonsense to speak about atoms as the current worldview does. Behind the sense perceptions, behind the sense qualities, [...] are not vibrations, but spiritual beingness. The world outside becomes ever more spiritual the further we advance in knowledge." [181]

When asking what matter is, Rudolf Steiner refers from the substance to the different chemical elements that appear through analysis. He says [...] "and that one could trace back the possibility of subdivision to a fundamental substance, which then through inner combinations, through the nature of its basic inner being, manifests at one time as gold, at another time as potassium, calcium. [...] There is a fundamental essence of our material Earth existence out of which all matter only comes into being by a condensing process. [...] Every substance upon the Earth is condensed light. There is nothing in material existence in any form whatever which is anything other than condensed light. [...] All matter is, in its essence, light." [182] It is interesting to note that modern electrodynamics only concluded decades later that all matter consists of some *quanta of energy*, of condensed light: *the photon is a field quantum or binding particle of the electromagnetic force* – from visible natural light to the condensed invisible "light" of X-rays and electron radiation below nature. If the light is further condensed, then electricity is created: "Then one recognizes that the atom can be nothing else than frozen electricity, frozen heat, frozen light. Then one has still to further to see the spirit in <u>all</u> that is condensed and formed. There is no matter! What matter is relates to the spirit as ice relates to water: if you dissolve the ice, there is water. If you dissolve matter, it disappears as matter and becomes spirit. Everything that is matter is spirit; matter is the external manifestation of the spirit." [183] That matter disappears is already shown by every chemical analysis, for in this process, the analyzed matter disappears, and something entirely different appears, namely the chemical elements.

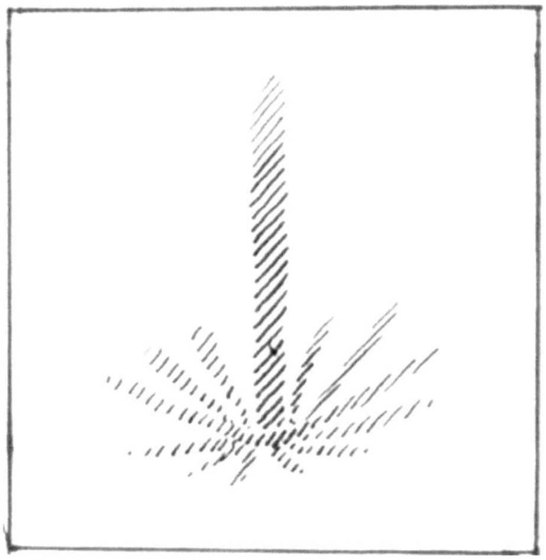

Fig. 100: Rudolf Steiner's sketch shows how spiritual form reaches a limit and then spreads and becomes a spatial form.

Rudolf Steiner's sketch in Fig. 100 shows how “spiritual form reaches a limit and then bursts and becomes a spatial form.” In a way, the spirit descends from the hierarchies and appears as a shattered form, as matter in the sense world. “The form comes from the supersensible with a specific movement, bursts, springs back into itself. [...] Matter is therefore in reality spirit, but shattered spirit. If now this work continues from will, wisdom, movement, form, even further beyond form; [...] namely, if a process in the universe has progressed up to the level of form, which is still entirely in the soul-spiritual realm, which is not yet a spatial form, [...] then the only possible next step is that the form as such breaks up, [...] then you have the transition from the supersensible into the sense-perceptible in space. And that which is shattered form is matter. [...] But before it shattered, it still had form both inwardly and spiritually. What was present as spirit form reverberates when it is shattered and in ruins. Where this has a strong after-effect, the lines of the spiritual forms continue, and there, after the piece shatters, an after-effect of the spiritual lines is still expressed in the lines which they describe. Thus crystals are formed. Crystals are reproductions of spiritual forms that still kept their original direction through their impetus but in the opposite sense. [184]

Becoming Substance

Pushed into the human senses
From depths of worlds as riddles
Is the fullness of matter.
Streaming into the depths of soul
From heights of worlds with meaning
Is the spirit’s clarifying word.
They meet within the human being
As wisdom-filled reality.

(Rudolf Steiner, CW 60)

Understanding Form

When man invented a world
Fragmented in endless atoms
His knowledge connected
With the death of nature;
He must now strive in spirit
To find what overcomes the fragments
And he will direct his knowing
To the evolution of the world.

(Rudolf Steiner, CW 263/I)

17. The Descent of the Nature Kingdoms in Human Evolution

Rudolf Steiner's research results have made it possible to overcome the limits of knowledge in the broad field of mineral substances and bring the phenomena into a meaningful context. A central idea of spiritual science for the Earth's development is the idea of a life sphere from which all the nature kingdoms have separated. All phenomena of the Earth sciences can be examined on this basis. For the sciences, this is new and unusual because they start from the current lifeless Earth and hypothesize that life is derived from chemical-physiological processes. The evolution of the Earth can be explained in spiritual scientific terms through an intuitively formulated concept of an etheric body, making it possible to understand the numerous phenomena of rock formations in context. All fluids in life are in a liquid, colloidal state. This phenomenon can still be observed today in bacterial life processes and material cycles, which means how substances are specifically enriched, bound, and excreted. These processes are still taking place today at the boundary between the living and the dead, which is taken as a phenomenological indication of the earlier life processes active in the living Earth organism and resulted in the separation of mineral substances. In numerous fields, this makes it possible to solve open questions about rock and mineral formation and create an overall picture that is mutually illuminating and supportive, which is not the case with current geoscientific models. It is only necessary to recall a few illustrations here, such as the differences in the emergence of the Earth with the other cosmic bodies, the relationship of the mica and feldspar sequences, plant and animal development, two different paths of coal formation, the solution to the agate problem, the formation of the numerous strangely embedded rock lenses, the cause of the rhythmic processes in all rock formations, to the shaping of global landscapes with their peneplains, high valleys, terraces, and notched valleys.

A new experience is created from the contemplation and thorough study of the phenomena. These phenomena are inwardly like an organism when applied to the whole Earth. In this way, the laws of the organic can be applied to the whole Earth. Rudolf Steiner shows how cause and effect in the sense-world can be observed and recognized in inorganic processes. One arrives at basic phenomena that lie within the experience and require no further explanation. However, actual knowledge of the organic is missing; there is no knowledge from which the processes of becoming, evolution, and devolution can be understood. From this follows the spiritual side: "that concept and reality which in the inorganic are one, are now separated. One can only call a concept that is not taken from the sense world by abstraction, but rather has a content flowing out of itself and only out of itself, an *intuitive concept* and knowledge of this concept an "intuitive" one. What follows from this is clear: *an organism can only be grasped in the intuitive concept*." [185] Rudolf Steiner gives a further explanation: "For the organic, it is necessary to allow one concept to grow out of another that in the progressive living metamorphosis of concepts there come to light images of that which appears in nature as a being possessing form. This Goethe strove to do in that he sought to hold fast in his mind an ideal image of a leaf which was not a fixed lifeless concept but such a one as might present itself in the most varied forms." [186]

The idea of the organism, which Goethe calls the "type" in plants, is even more real than any single organism because it is revealed in every organism. This type appears as a creative force and brings about the metamorphoses which appear side by side in the sense world. Similarly, Rudolf Steiner chooses the comparison with a tree trunk to understand the whole Earth with all the kingdoms of nature as one living thing, which has arisen from an overall life, in contrast to the otherwise materialistic ideas. "[...] Imagine the Earth [when] speaking of a tree and fixing one's gaze only upon the trunk, without its leaves, blossoms, and fruit. To the sight of our physical eyes, such a tree trunk can be called reality. But in a higher sense, it is not reality. For, as a mere trunk, it cannot occur as such anywhere in our world. It can be what it is only in so far as those growth forces arise in it simultaneously, which unfolds the leaves, blossoms, and fruits. One must think about the trunk's reality and include these forces, and be conscious that the trunk by itself gives the beholder a deceptive picture of reality. The fact that something or other is present to the senses is not yet proof that it is also a reality in this form. [...] Whoever wants to picture something real about the Earth must picture it in such a way that its mineral realm already contains within itself the plant realm, just as the trunk configuration of the tree includes its leaves and blossoms; yes, that within the "true Earth" the animal realm and human being are already present along with it. [...] Certainly, everyone pictures the Earth as a planetary body with plants, animals, and human beings. [...] But this Earth as a sum, added up out of minerals, plants, animals, and human beings, does not exist at all. It is only a delusion of the senses. [...] There is a true Earth; it is an entirely supersensible configuration, an invisible being, which provides the mineral foundation from out of itself; but it is not limited to this, for it manifests itself further in the plant realm, then in the animal realm, then in the human realm." [187]

Rudolf Steiner drew the Earth in the Hyperborean period as a white circle in which the granite layer had already become somewhat solidified over the basaltic (Fig. 101). Throughout the entire cosmos, only on Earth do we find the granitic. All solid bodies in space consist of the basaltic, including the planetoids, Mars, Venus, Mercury, the Moon, and the Earth under the continents and oceans. Rudolf Steiner describes this stage of development in a public lecture: "As we go further back our Earth appears more and more like an utterly different planetary body, becomes something different; [...] with the character of a living being. [...] We have within the surrounding of our Earth [...] something like living members of a great living being. [...] Then we come to a time when we must say that there is such a high temperature on the Earth's surface that life cannot develop, but in its surroundings, life develops which wants to descend but cannot. Why is it that it cannot descend? Through the fire-process, down in the depths, life cannot develop under conditions of intense heat. [...] Later on, it separated from our Earth in the same way that our living organism separated the solid parts and the bones from the soft parts. And now we look at the granite-formation and say the material which the granite contains — quartz, feldspar, and mica – is initially in a state of dissolution within the great living being of the Earth. The latter needs to discard these substances for its development: it exudes them and lets them fall to the Earth. What is below absorbs these exudations, builds up a basic massif, a skeleton in the living Earth being. [...] And when the Earth approached the time when granite and similar material segregated, it was laid hold of by all such destructive processes — which were the means whereby forces of spirit-and-soul worked upon matter. Into that organism into which formerly were worked not only the substances appertaining today to the plant, animal, and human organisms but also the substances which today constitute our Earth-massif – into this organism flowed the effects of all these destructive processes that were due to the working of forces of spirit and soul. These destructive processes introduced into the great living Earth being that which has been discarded by an elimination process, and what we have today as mineral substances that build up our Earth today, which we do not find in organic bodies." [188]

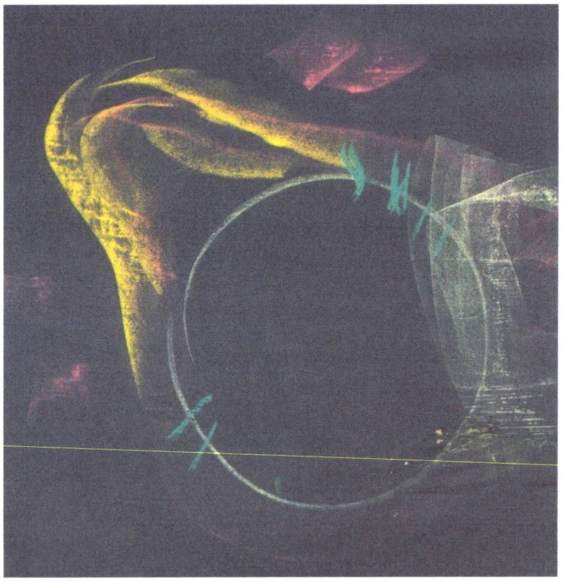

Fig. 101: In a lecture, Rudolf Steiner drew the Earth in the Hyperborean period as a white circle in which the granite layer was solidifying over the basaltic.

Rudolf Steiner describes the mineral as a dead state no longer controlled by an etheric body's life forces. With the death of an organism, the mineral portion detaches itself, and the physical-chemical processes of weathering begin to decompose the rocks. Compared to its full size, today's Earth only has a very thin solid layer, and it is dead. This condition also true for the physically visible part of the water and air. They are still permeated by warmth and the formative forces of their ether body. This process is exemplified in Fig. 101, which shows how high up in the hot air layers of the former life sphere the etheric works, how the Earth's mineral substances condensed, and then in a liquid state sank, forming the solidifying Earth surface. The soft-bodied plants already touched the Earth in the Hyperborean epoch (this is indicated in Fig. 101 as blue-green lines). Graphite and bituminous coals come about from these plant-like substances in the layers of shale and gneiss. All that is animal, indicated in the reddish areas, was still in the life sphere's upper regions. It only came down to the Earth after the Earth-Moon body had separated from the spheres of the planets and the Sun in the 3rd and 4th Lemurian epoch, in the

Paleozoic era, the white area on the right. However, everything was affected by and separated from human development, indicated by the large yellow arcs.

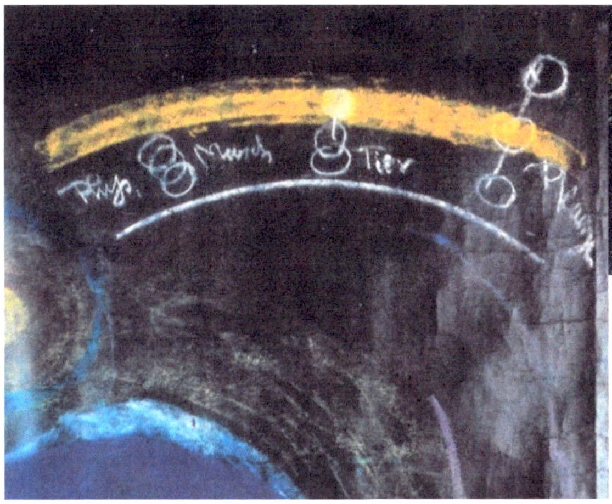

Fig. 102: Rudolf Steiner Blackboard Drawing

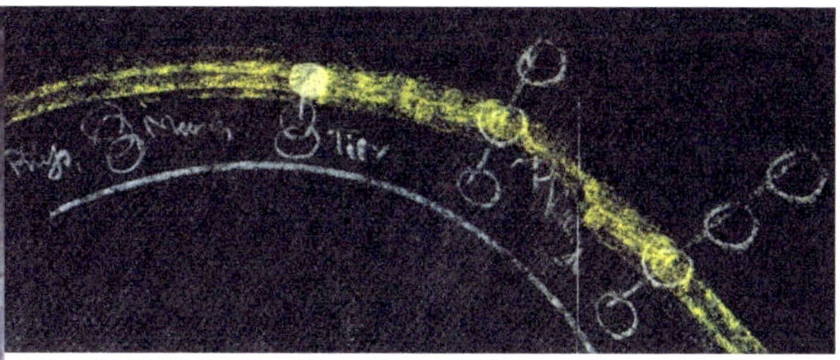

Fig. 103: Dankmar Bosse's extrapolation of Fig. 102 shows the mineral kingdom, added on the right side.

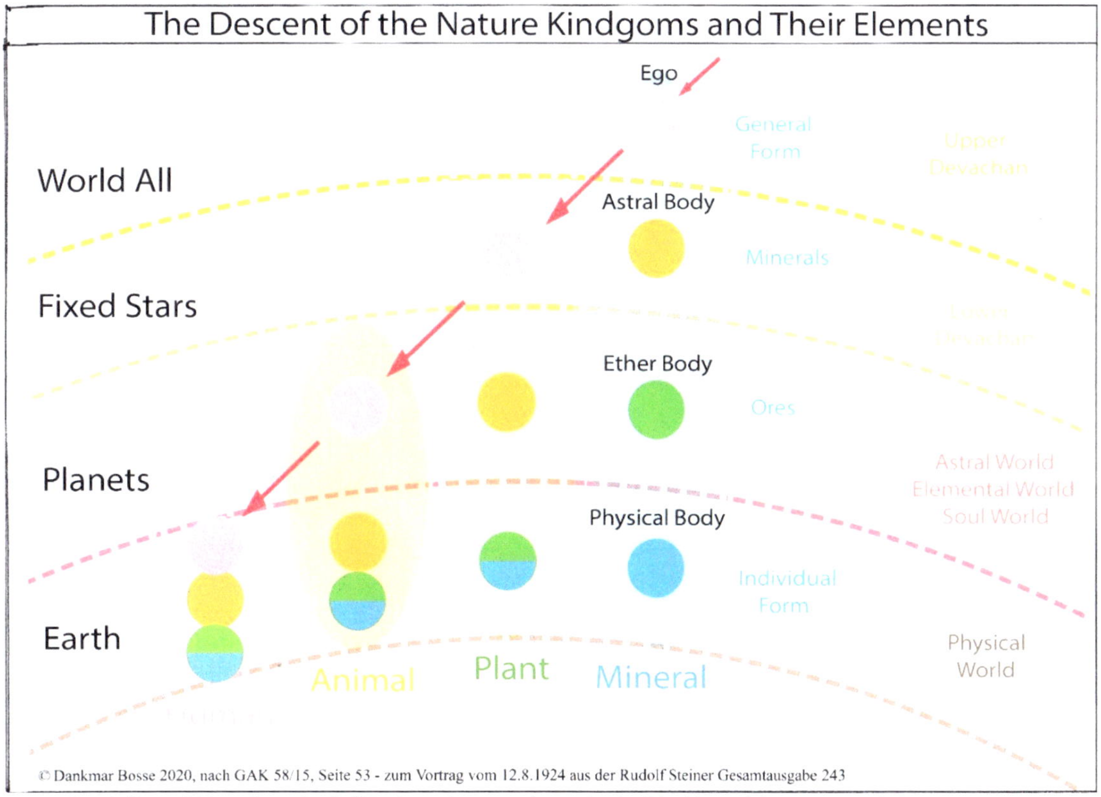

Fig. 104: Collected ideas from Rudolf Steiner that are shown in the blackboard drawings and elsewhere.

Fig. 102 is a blackboard drawing that Rudolf Steiner drew during a lecture in Torquay. The white arc shows the Earth, and above it is the physical world where the human, animal, and plant have been identified. The next layer is the astral, or elemental world, which is shown in yellow. The three circles in a line to the right point to the physical and etheric body below, above them the astral body, on top is the "I," which lives in the physical-etheric and the lower spirit world. The word plant was written diagonally because the blackboard ended. You can perhaps understand that the geologist in me was missing the mineral kingdom present within the Earth's white circle. So I took the liberty of adding the mineral kingdom to the picture, as shown in Fig. 103. The "I" is in the higher spirit world that fills the world all. The assembled results can be seen in Fig. 104.

Figure 105 is an example of a group of rock salt crystals, where each crystal has the exact form of a cube created from the formative forces of the world all, but in which each crystal has a form brought about by Earth forces. Figure 106 shows a group of minerals in a typical granite specimen whose forms originate from the lower spirit world of the fixed stars; in this case: black mica, red and white feldspar, and grey quartz. Figure 107 shows a cluster of minerals between white rock crystals. In this example, the mineral is pyrite, whose forms and shapes were developed in the same way as the rock salt crystals. However, the substance was irradiated into the granitic in subtle quantities from the elemental world of planetary forces, from which it was then concentrated and deposited in ore veins. Fig. 108 shows a group of plant leaves as they metamorphose from the rosette via the lower and upper leaves to the sepals, petals, stamens, and pistil. Goethe wrote about this: "Forwards and backward the plant is always only leaf." And in Fig. 109, ostrich birds illustrate how most animals have a group ego in common whereby the single animal is individually formed. Thus it becomes understandable that plants can be bred and grafted within a family, which is impossible for animals.

Fig. 105: Rock Salt Crystals

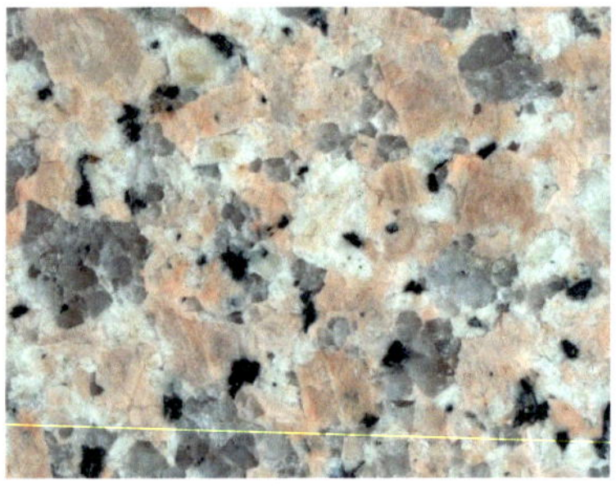

Fig. 106: Minerals in a typical granite sample.

Fig. 107: Pyrite and quartz crystals.

Rudolf Steiner describes the descent of the kingdom of nature from the development of the human being in Fig. 102: "And now we discover that, as human beings, we bear within us that which is the real being of the plants in the higher spheres. We bring down into this mirrored life what the plants leave behind in the world of spirit, for the plant-beings live in that world and send down to Earth their reflected images, and the Earth fills them with earthly substance. We human beings bring our soul-nature, which also belongs to that higher world, into this world of images. We are not mere images, but we are also spiritual beings of soul here on Earth. On Earth, we participate in three worlds. We live in the physical world, where animals' self-consciousness is not to be found; simultaneously, we inhabit the astral world where their self-consciousness exists,

Fig. 108: Ostriches are all individual copies of a specific group form.

Fig. 109: Plant Rosette.

and this astral world we bring down into the physical world. We also inhabit a third world, the spiritual world, where dwell the true plant-beings, but the plant-beings send only their reflected images down to Earth, whereas we bring down the realities of our soul-life. And now we can say: a being who possesses body, soul, and spirit here on Earth is a human being. A being with body and soul here on Earth, but whose spirit dwells in a second world bordering on the physical world and which for that reason has less reality, is an animal. A being with only a body in the physical world, the soul in the second world, and the spirit in the third world, so that the body is only a reflected image of the spirit and is filled out with terrestrial matter, a plant. We now understand the three worlds in nature, and we know the human being bears these three worlds within. We feel the plants reaching up to the stars to some extent. As we look at the plants, we say to ourselves: here is a being that manifests only its reflected image on Earth, an image detached from its true reality. The more we direct our gaze to the stars at night, the more do we see its true being in the higher worlds. When we look from Earth to heaven and perceive the cosmos to be one with the Earth, then we see the world of nature as a totality. Then we look back at ourselves as human beings and say: whatever reaches up to there in the plant, we have brought together in ourselves on Earth. We bear within ourselves the physical, astral, and spiritual worlds." [189]

Author Biography

DANKMAR BOSSE was born in Weimar, Germany, in 1940. He studied whole earth geology, mineralogy, zoology, and botany at the Freiberg University of Mining and Technology in Saxony. He worked for twenty-one years at the Central Geological Institute in East Berlin on regional geology and middle European tectonics. During this time, he tried to relate natural science facts to Rudolf Steiner's research results, as described in Chapter 15, The Earth was a Living Being. Since then, he has been independently active with seminars, lectures, excursions, and anthroposophic research work in all Earth-scientific fields and numerous research expeditions.

He is the editor of the book J. W Goethe: Die Metamorphose des Granits (J. W Goethe: The Metamorphoses of Granite), an edited collection of Goethe's geologic writings, published by Verlag Freies Geistesleben in 1994, and the German edition of Die Gemeinsame Evolution von Erde und Mensch, Entwurf einer Geologic und Paliiontologie der lebendigen Erde (The Mutual Evolution of Earth and Humanity: Sketch of a Geology and Paleontology of the Living Earth) in 2002 by the same publisher (2019 by SteinerBooks in English). In 2015 his comprehensive book on mineralogy from a natural-science/anthroposophic viewpoint -Die Evolution der Minerale zwischen Kosmos und Erde (The Evolution of the Minerals between Cosmos and Earth, Sketch of a Mineralogy and Crystallography of the Living Earth) was published by the Verlag des Ita Wegman Instituts, Stuttgart.

References

[1] Rustemeyer, Paul, *Tourmaline: Fascinating Crystals with Fantastic Inner Worlds*, Verlag Dr. Friedrich Pfeil, German-English, 2015).

[2] Steiner, R. *Stones and Metals in Human Evolution*, October 13, 1906, CW 97.

[3] Steiner, R. *Esoteric Development*, Lecture IX: Imaginative Knowledge and Artistic Imagination, October 21, 1906.

[4] Steiner, R. (1906). Skizzen und Entwürfe zur Kleinodienkunst. Gesamtausgabe Nr. K51 Dornach; p. 96 – 102, p. 311 – 313.

[5] Lüschen, H. (1979): *Die Namen der Steine*. Thun.

[6] Giegner, Michael, for example, The Healing Crystals First Aid Manual and Crystal Power, Earthdancer Books, 2006; and, Crystal Healing: The Complete Handbook, Cassel, 2015.

[7] Steiner, R. *Course for Young Doctors*, Lecture VII, January 8, 1924, CW 316.

[8] Ibid.

[9] Steiner, R. *Spiritual Science and Medicine*, Lecture XI, March 31, 1920, CW 312.

[10] Steiner, R. *An Outline of Occult Science*, pp. 102-240, Anthroposophic Press, New York, 1922, CW 13.

[11] Steiner, R. *Spiritual Science and Medicine*, Lecture XI, March 31, 1920, CW 312.

[12] Steiner, R. *Origins of Natural Science*, Lecture IX, January 6, 1923, CW 326.

[13] Steiner, R. Anthroposophical Spiritual Science and Medical Therapy, Lecture IX, April 18, 1921, CW 313.

[14]Bosse, Dankmar. The Mutual Evolution of Earth and Humanity: Sketch of a Geology and Paleontology of the Living Earth (p. 259). SteinerBooks. Kindle Edition.

[15] Ibid. (p. 261). SteinerBooks. Kindle Edition.

[16] Ibid, (inside of back cover).

[17] Steiner, R. World History, Lecture 5: Mysteries of the Est, West, and of Ephesus, Dec.r 28, 1923, CW 233.

[18] Bosse, Dankmar. The Mutual Evolution of Earth and Humanity: Sketch of a Geology and Paleontology of the Living Earth (Fig. 177, p. 219, and Fig. 168, p. 211). SteinerBooks.

[19] Bosse, Dankmar. The Mutual Evolution of Earth and Humanity: Sketch of a Geology and Paleontology of the Living Earth (p. 259). SteinerBooks. Kindle Edition.

[20] Ibid. (p. 261). SteinerBooks. Kindle Edition.

[21] Steiner, R. Spiritual Science and Medicine, Lecture XI, March 31, 1920, CW 312.

[22] Ibid., Lecture X, March 30, 1920.

[23] Steiner, R. *An Occult Physiology*, Lecture 2, March 21, 1911, CW 128

[24] Steiner, R. Spiritual Science and Medicine, 1920, CW 312

[25] Steiner, R. CW 323, p. 148, German edition).

[26] Steiner, R. *Spiritual Science and Medicine*, Lecture IX, March 29, 1920, CW 312

[27] Steiner, R. *Spiritual Science and Medicine*, 1920, CW 312 and CW K 58/22, p. 25 in the German version

[28] Originally named "cinnamon stone" ("Kanelstein" in German) in 1803 by Abraham Gottlob Werner and renamed grossularite by Werner in 1808. Named for the color of gooseberries (Ribes grossularium) as the original specimens were this color.

[29] Steiner, R. *Spiritual Science and Medicine*, Lecture IX, March 29, 1920, CW 312

[30] Ibid.

[31] Steiner, R. *Spiritual Science and Medicine*, Lecture XIX, April 8, 1920.

[32] Ibid., Lecture XII, April 1, 1920.

[33] Ibid., Lecture XX, April 9, 1920.

[34] 

[35] Manfred Schmidt-Brabant, *Michael- Gedanken und Drachen- Kräfte. Anregungen zu einem künftigen*, Verlag am Goetheanum, October 1, 1992. Not available in English.

[36] Goethe, J.W, Zahme Xenien 2: 652-653; Dummes Zeug kann man viel reden,/Kann es auch schreiben,/Wird weder Leib noch Seele töten,/Es wird alles beim alten bleiben./Dummes aber, vors Auge gestellt,/Hat ein magisches Recht;/Weil es die Sinne gefesselt hält,/Bleibt der Geist ein Knecht.

[37] Revelation of John, 12:7

[38] Ibid. 12:14

[39] Luis Walter Alvarez, Jun 13, 1911 - Sep 01, 1988 (age 77), was an American experimental physicist, inventor, and professor who was awarded the Nobel Prize in Physics in 1968 for the development of the hydrogen bubble chamber enabling discovery of resonance states in particle physics.

[40] Alvarez LW, Alvarez W, Asaro F, Michel HV. Extraterrestrial cause for the cretaceous-tertiary extinction. Science. 1980;208(4448):1095-1108. doi:10.1126/science.208.4448.1095

[41] The **Alvarez hypothesis** posits that the mass extinction of the dinosaurs and many other living things during the Cretaceous–Paleogene extinction event was caused by the impact of a large asteroid on the Earth. Prior to 2013, it was commonly cited as having happened about 65 million years ago, but Renne and colleagues (2013) gave an updated value of 66 million years. Evidence indicates that the asteroid fell in the Yucatán Peninsula, at Chicxulub, Mexico. See https://en.wikipedia.org/wiki/Alvarez_hypothesis.

[42] Bosse, Dankmar. The Mutual Evolution of Earth and Humanity: Sketch of a Geology and Paleontology of the Living Earth, (pp. 106-107), SteinerBooks, 2019.

[43] H. Jäger, Zs. Geol. Wiss. 14, H.6. 1986; or D. Thies, Geowiss. 9, No. 2, 1991

[44] Prof. Dr. Detlev Thies, Leibniz Universität Hannover, Institut für Geologie, Callinstraße 30 30167 Hannover

[45] Steiner, R. *Man's Fall and Redemption*, January 26, 1923, CW220.

[46] Steiner, R. Spiritual Science and the Art of Healing, July 17, 1924, CW 319.

[47] Bosse, Dankmar. *The Mutual Evolution of Earth and Humanity: Sketch of a Geology and Paleontology of the Living Earth*, Spiritual Evolution, (pp. 116-19) and (pp. 73—128, among others.), SteinerBooks, 2019.

[48] Stanley Miller, American chemist (born March 7, 1930, Oakland, Calif.—died May 20, 2007, National City, Calif.), designed the first experiment to produce organic molecules from some of the inorganic components of the Earth's prebiotic atmosphere. Miller's procedure (which was co-designed by Harold Urey and is known as the Miller-Urey experiment) contained three key elements: a heated pool of water meant to simulate the primitive Earth's ocean; an atmosphere of water vapor, methane, ammonia, and molecular hydrogen; and storms of "lightning" in the form of continuous electric discharges. After one week, 10–15% of the system's carbon was found in organic molecules such as amino acids, the building blocks of proteins. Miller's experiment was not only a groundbreaking moment for research into the origin of life on Earth but also a breakthrough that captured the popular imagination and gave rise to the term prebiotic soup. (Online citation from Encyclopedia Britannica).

[49] Steiner, R., *The Philosophy of Freedom: The Basis for a Modern World Conception*, Rudolf Steiner Press, London, 1964, CW 4. Translation by Michael Wilson.

[50] Ibid.

[51] Steiner, R., *Anthropsophical Leading Thoughts*: The Way of Michael and What Preceded It. (pp. 68-69). Rudolf Steiner Press. Kindle Edition, CW 26

[52] Steiner, R., Notebook Archive - No. 314, for T*rue and False Paths in Spiritual Investigation*, Lecture Seven, August 18, 1924. CW 243

[53] Steiner, R., *Occult Science and Occult Development*: Christ at the Time of the Mystery of Golgotha and Christ in the Twentieth Century, May 2, 1913. CW 152

[54] Steiner, R., *The Festivals and their Meaning*, Lecture 5, The Michael Impulse and the Mystery of Golgotha, May 18, 1913. CW 152

[55] Steiner, R., *The Fall of Spiritual Darkness*, Lecture 9, The Battle between Michael and The Dragon, October 14, 1917. CW 177

[56] Steiner, R., *Occult Science and Occult Development*, May 2, 1913. CW 152

[57] Steiner, R., *From the Contents of the Esoteric Classes*, Munich, December 5, 1907. Also, see Contribution to CW 67/68, 1979. CW 68

[58] Steiner, R. *The Impulse for Renewal in Culture and Science*, Lecture One, Anthroposophy and Natural Science, March 6, 1922. CW 81

[59] Steiner, R., Thinking and Willing as Two Poles of the Human Soul-Life, July 15, 1921. CW 205

[60] Steiner, R., October 14,1917. CW 177

[61] Steiner, R, Haeckel, *The Riddles of the World and Anthroposophy*, October 5, 1905. CW 54

[62] Steiner, Rudolf. *Anthroposophical Leading Thoughts*: Anthroposophy as a Path of Knowledge - The Michael Mystery (pp. 127-128). Rudolf Steiner Press. Kindle Edition. CW 26

[63] Steiner, R., *Anthroposophical Leading Thoughts:* Third Study-Michael's Suffering over Human Evolution before the Time of his Earthly Activity. (p. 126). Rudolf Steiner Press. Kindle Edition. CW 26

[64] Steiner, R., *The Younger Generation*, Lecture XIII, October 15, 1922. CW 217

[65] Steiner, R., *Social Life*, Lecture 1, January 21, 1921. CW 203

[66] Steiner, R., *Behind the Scenes of External Happenings*, November 6, 1917. CW 178

[67] Steiner, R., October 14, 1917. CW 177

[68] Steiner, R., *The Mission of the Archangel Michael*, November 30, 1919. CW 194

[69] Steiner, R., *An Outline of Occult* Science, Anthroposophic Press, Inc., Spring Valley, NY, 1972.

[70] Ibid. p. xxix.

[71] Steiner, R., *Autobiography: Chapters in the Course of my Life*, Anthroposophic Press, 1999, p. 79. CW 28

[72] Goethe, Joahann Wolfgang, Goethe Complete Works (GCW), vol. 8, p.186 in the German Edition.

[73] Ibid. (GCW) Vol. 8, p. 74. in the German Edition.

[74] Goethe, Johann Wolfgang, Ed., Letters, July 13, 1816

[75] Goethe, Joahann Wolfgang, , Vol 1, Leopoldina Edition, p. 142 in the German Edition

[76] Goethe, WA, Letters, 29.5.1801

[77] Goethe, Joahann Wolfgang, Goethe Complete Works, Vol. 8, Leopoldina Edition, p. 164 in the German ed.

[78] Goethe, Joahann Wolfgang, Goethe Complete Works, Vol. 9, Leopoldina Edition, p. 307 in the German ed.

[79] Steiner, R., Autobiography: Chapters in the Course of my Life, Anthroposophic Press, 1999, pp. 111-112 CW 28

[80] Steiner, R., Steiner, R. What Has Geology to Say About the Origin of the World?, Berlin, February 9, 1911, CW 60

[81] Steiner, R., *An Outline of Occult* Science, Anthroposophic Press, Inc., Spring Valley, NY, 1972. Translated by Maud and Henry B. Monges and revised by Lisa D. Monges. CW 13

[82] Rudolf Steiner: *An Outline of Esoteric Science*, Afterword to the Small Edition of Works, Rudolf Steiner Publishing House, (1996) CW 13

[83] Steiner, R., Steiner, R. *What Has Geology to Say About the Origin of the World?*, Berlin, Feb 9, 1911, CW 60

[84] Cloos, Walther, *The Living Earth*, Lanthorne Press, 1977, p. 26

[85] In Hebrew tradition (Kabbalah), Adam Kadmon is the supreme essence, the archetype of humankind.

[86] Steiner, R., *Origin and Goal of the Human* Being, Lecture XI, February 9, 1905, CW 53

[87] Steiner, R., *Evolution, Earth, Man*, Lecture II, July 3, 1924, CW 354.

[88] Steiner, R., *What Has Geology to Say About the Origin of the World?*, Lecture XI, February 9, 1911. CW 60

[89] Steiner, R., *Lectures for the Workmen at the Goetheanum*: The Recognition of the Nature of the Human Being as Body, Soul and Spirit. On Earlier Conditions of the Earth; September 1922, CW 347.

[90] Steiner, R., *Universe Earth and Man*, Lecture VII, August 11, 1908, CW 105.

[91] Ibid.

[92] Steiner, R., *The Mission of the Individual Folk-Souls*: The mission of individual peoples and cultures in the past, present and future, Lecture Ten, June 16, 1910, CW 121.

[93] (Goethe, LA 8, p. 29)

[94] (Goethe, LA 8, p. 381-382).

[95] Bosse, Dankmar, *Die Evolution der Minerale zwischen Kosmos und Erde*, Stuttgart, 2015, Vol. 2, pp. 41-467

[96] Steiner, Rudolf, *The Cosmic Prehistoric Ages of Mankind*, Lecture III: Romanism and Freemasonry, September 9, 1918, CW 184.

[97] Bentley, W. A., *Snowflakes in Photographs*, Dover Pictorial Archive, 2012

[98] Goethe, J.W., *Writings on Natural Science*, Weimar Edition (German), 1897-1919, Section II: V. 10, S. 256)

[99] Steiner, R., *The Four Seasons and the Archangels*, Lecture I: The Michael Imagination. The Michael Imagination verse was translated by Arvia MacKaye Ege. CW 229

[100] Steiner, R. and Wegman, I., Fundamentals of Therapy: An Extension of the Art of Healing through Spiritual Knowledge. CW 27

[101] Steiner, R., The Spiritual Beings in the Heavenly Bodies and in the Kingdoms of Nature, Lecture 6, April 8, 1912. CW 136

[102] ibid, Lecture 9, April 13, 1912

[103] Steiner, R., The Influence of Spiritual Beings on Man, Lecture 3, February 15, 1908. The author took the liberty of adding the boundaries of the spheres, the planetoid belt and the ages. The sizes and distances cannot be represented to scale in a drawing. CW 102

[104] Steiner, R., The Spiritual Hierarchies and Their Reflection in the Physical World, Lecture 5, April 14, 1909. CW 110

[105] Steiner, R., Man as Symphony of the Creative Word, Evolution of Animals and Man, Lecture 6, October 28, 1923. GA K 58/13, p. 51. CW 230

[106] Steiner, R., Theosophical/Esoteric Cosmology. CW 89

[107] Steiner, R., The Apocalypse of St. John, Lecture X, June 27, 1908; GA 104

[108] Steiner, R., The Gospel of John, Lecture IV, June 27, 1909, CW 112

[109] Steiner, R., August 2,1908, CW 9

[110] Steiner, R., October 28,1905 CW 93a

[111] Steiner, R., Esoteric Cosmology: Lecture X, 1906-06-06. CW 94

[112] Steiner, R., Mystery Centers: Lecture IV: The Ephesian Mysteries of Artemis, November 30, 1923; Lecture 5, December 1, 1923, CW 232; Lecture 11, The Secret of Plants, of Metals, and of Men. CW 313: *Anthroposophical Spiritual Science and Medical Therapy*: Lecture VIII, April 18, 1921. CW 316: Steiner, R., Young Doctor's Course, Lecture 3, April 1, 1924.

[113] Steiner, R., Theosophy of the Rosicrucian, The Technique of Karma, May 31, 1907 CW 99

[114] Steiner, R., Chapter VII, CW 27

[115] Steiner, R., Art of Healing, Lecture III, July 24, 1924. CW 319

[116] Steiner, R., The Annual Cycle as the Breathing Process of the Four Great Feast Days. Lecture 5, April 4, 1923. CW 223

[117] Steiner, R., September 3, 1924, CW 319

[118] Steiner, R., December 26, 1923, CW 233

[119] Steiner, R., CW 213, July 2, 1922, CW 232, December 1, 1923, CW 312, March 30, 1920)

[120] Steiner, R., Paths to Lost Forces Working in Nature, "Die Erkenntnis-Aufgabe der Jugend" CW 217a

[121] Steiner, R., Lecture 4, June 12, 1924, CW 327

[122] Steiner, R. Agriculture Course: The Birth of the Biodynamic Method (Classic Translation), Rudolf Steiner Press, 1958, Translated by George Adams, Kindle Edition. CW 327: Lecture IV, Page 75; Supplement, p. 159.

[123] Mackensen, Manfred v. (1994): Process Chemistry. Kassel.

[124] Steiner, R., 1910

[125] Ibid.

[126] Bosse 2002

[127] Steiner, R., The Agricultural Course, June 7-16, 1924, Translation by George Adams, Preface by Ehrenfried Pfeiffer.

[128] Ibid.

[129] Ibid.

[130] Ibid.

[131] Further aspects can be found in Rudolf Steiner's notebooks Steiner 1967, p. 13

[132] Further aspects can be found in Rudolf Steiner's notebooks (Steiner 1967, p.16):

[133] Steiner, R. (1967): Rudolf Steiner's notes on the agricultural course. Contributions to the Rudolf Steiner Complete Edition No. 18 Dornach. http://fvn-archiv.net/PDF/Beitraege/BE-018-1967.pdf

[134] Remer, Nikolaus (1980): Organic Manure: Its Treatment According to Indications by Rudolf Steiner. Wynstones Press.

[135] Pfeiffer, Martin W. (1975): Martin W. Pfeiffer, Die lanwirtschaftliche Individualitat - Ein Bild des Menschen Zum Verstandnis der Praparate der biologisch-dynamischen Wirtschaftweise, Selbstverlag Borchen, 1975. Self-published.

[136] Bosse, Dankmar. *The Mutual Evolution of Earth and Humanity: Sketch of a Geology and Paleontology of the Living Earth*, End Piece, SteinerBooks, 2019

[137] Stappung, W. (2005): Hornton, ein zusätzliches Präparat? Lebendige Erde H. 1, S. 44-45. Darmstadt.

[138] (Steiner 1910; Bosse 2002)

[139] Steiner, R., I. Wegman *Fundamentals of Therapy*, Chapter VII, Dornach, 1925. CW 27

[140] Steiner, R., The Agricultural Course, June 7-16, 1924, Translation by George Adams, Preface by Ehrenfried Pfeiffer. Also, Steiner, R., and Wegman, Ita, *Fundamentals of Therapy*, Chapter VII, CW 27

[141] Steiner, R., The Agricultural Course, June 7-16, 1924, Translation by George Adams, Preface by Ehrenfried Pfeiffer.

[142] Ibid.

[143] Ibid.

[144] Blöss, C., H.-U. Niemitz (1997): Der C14-Crash. Gräfelfing (D). www:efodon.de

[145] Schweingruber, 1983

[146]: Steiner, R., An Outline of Occult Science, 1910, CW 13

[147] 1882 ("The Only Possible Criticism of Atomistic Concepts"- Contributions to CW 63, 1978), and again in 1890 CW 1, Chapter XVI/2

[148] Steiner, R., February 4, 1913, CW 262

[149] Geyh, M. A. (1983): Konfusion über quartäre Zeitskalen. Geol. Jahrbuch, A 71, Stuttgart. Also, Blöss, C., H.-U. Niemitz (1997): Der C14-Crash. Gräfelfing (D). www:efodon.de

[150] Physical Review Letters Nr. 93/2004

[151] Steiner, R., March 13, 1920, CW 321

[152] Steiner, R., January 6, 1923, CW 326

[153] Steiner, R., July 30, 1920, CW 197

[154] Steiner, R., October 5, 1905, CW 93a

[155] Steiner, R., March 9, 1905, CW 53

[156] Steiner, R., September 20, 1918, CW 184

[157] Steiner, R., December 31, 1915, CW 165
[158] Steiner, R., September 17, 1910, CW 125
[159] Steiner, R.,, November 12, 1917, CW 73
[160] Bosse, Dankmar. The Mutual Evolution of Earth and Humanity: Sketch of a Geology and Paleontology of the Living Earth. See Separation of the Moon from the Earth, SteinerBooks.
[161] In Kniebe, G.: What is time? Stuttgart, 1993
[162] Wachmuth, 1950
[163] Bosse, Dankmar. The Mutual Evolution of Earth and Humanity: Sketch of a Geology and Paleontology of the Living Earth. SteinerBooks.
[164] Steiner, Rudolf, *The Influence of Spiritual Beings on Man*, Lecture III, February 15,1908, CW 102.
[165] Bosse, Dankmar. The Mutual Evolution of Earth and Humanity: Sketch of a Geology and Paleontology of the Living Earth. See *Separation of the Moon from the Earth* p. 284. 2019, SteinerBooks.
[166] Steiner, Rudolf, *An Outline of Occult Science*, Anthroposophic Press, 1972.
[167] Bosse, Dankmar. The Mutual Evolution of Earth and Humanity: Sketch of a Geology and Paleontology of the Living Earth (The Formation of Landscapes Out of the Plain, p. 343). SteinerBooks.
[168] Steiner, Rudolf, What Has Geology to Say About the Origin of the World? February 9, 1911, CW 60.
[169] Rösler, H.J. (1981): Lehrbuch der Mineralogie. Leipzig.
[170] Raaz, Franz & Köhler, Alexander. Bau und Bildung der Kristalle: die Architektonik der stofflichen Welt, Springer, 1953.
[171] Rustemeyer, Paul (2003) Faszination Turmalin: Formen - Farben - Strukturen
[172] Borchardt-Ott, Walter (2011) Crystallography: An Introduction
[173] Rustemeyer, Paul (2003) Footnote 3.
[174] Rykart, Rudolf (1989) Quarz-Monographie die Eigenheiten von Bergkristall, Rauchquarz, Amethyst, und anderen Varietäten
[175] Ibid.
[176] Bienlein, 2003
[177] Steiner, Rudolf. *Anthroposophical Leading Thoughts: Anthroposophy as a Path of Knowledge* - The Michael Mystery (p. 69). Rudolf Steiner Press. Kindle Edition.
[178] Baumgartner, Stefan 2014
[179] Steiner, Rudolf. *Goethean Science*, Lecture XV, 1884, CW 1
[180] Steiner, Rudolf. October 6, 1920, Supplement to CW Nr.114/115
[181] Steiner, Rudolf. Psychoanalysis in the Light of Anthroposophy, Connections Between Organic Processes and the Mental Life of Man, CW 205, July 2, 1921
[182] Steiner, Rudolf. *Manifestations of Karma*, Lecture 10, CW 120, May 27, 1910
[183] Steiner, Rudolf. *Knowledge of Soul and Spirit.* CW 56, October 17, 1907
[184] Steiner, Rudolf. *The World of the Senses and the World of the Spirit*, Lectures IV and V, CW 134, December 30 and 31, 1911
[185] Steiner, R. *Goethean Science*, The Mercury Press, 1988, CW 1. Translated by William Lindeman.
[186] Steiner, R. *The Story of My Life*, Chapter VI, Anthroposophical Publishing Co. (London), 1928, CW 28.
[187] Steiner, R. The Riddle of Man: A Forgotten Stream in German Spiritual Life, 1916, CW 20.
[188] Steiner, R. What Has Geology to Say About the Origin of the World?, Berlin, February 9, 1911, CW 60
[189] Steiner, R. *True and False Paths in Spiritual Investigation*, Lecture 2, August 12, 1924, Torquay, CW 243.

Made in the USA
Las Vegas, NV
04 November 2026

57423059R00090